T0210181

ZEPPELIN!

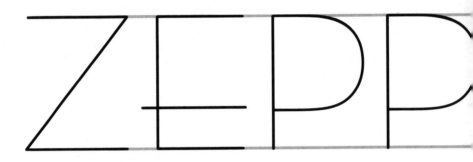

THE JOHNS HOPKINS UNIVERSITY PRESS · BALTIMORE & LONDON

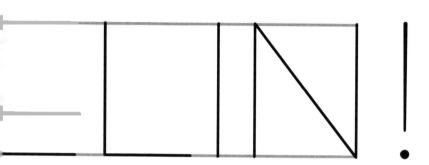

ermany and the Airship, 1900–1939

Guillaume de Syon

© 2002 The Johns Hopkins University Press
All rights reserved. Published 2002
Printed in the United States of America on acid-free paper

9 8 7 6 5 4 3 2 1

The Johns Hopkins University Press
2715 North Charles Street
Baltimore, Maryland 21218-4363
www.press.jhu.edu

Library of Congress Cataloging-in-Publication Data

De Syon, Guillaume, 1966–
   Zeppelin! : Germany and the airship, 1900–1939 / Guillaume de
Syon.
      p. cm.
Includes bibliographical references and index.
   ISBN 0-8018-6734-7 (hardcover)
   1. Airships—Germany—History. 2. Zeppelin, Ferdinand, Graf
von, 1838–1917. I. Title.
   TL658.Z4 .D33 2001
   629.133′25′0943—dc21

2001000240

A catalog record for this book is available from the British Library.

A la mémoire de Papa

CONTENTS

ACKNOWLEDGMENTS

This book owes much to the patience and support of the many individuals who took the time to read, comment on, and considerably improve its drafts. First and foremost I would like to thank my doctoral advisor, Dietrich Orlow, without whom this project could not have been either contemplated or completed. In addition to intellectual stimulation, Dietrich and his wife Maria offered friendship and hospitality in Boston and Austria. Their encouragement and aid were extraordinary. I am also grateful to Boston University's Lancelot Farrar, Thomas Glick, James Johnson, William Keylor, and Robert Schulmann; their assistance in the last stages of completion of the dissertation was crucial.

Throughout the course of my research, staff members at the numerous archives and resource centers I visited were unfailingly patient and helpful. I would like to mention especially Frau Ganser, Frau Jakobi, and Herr Scharmann at the German Federal Archives in Koblenz; Klaus Dietrich and Frau Wolf at the German Air and Space Society in Bonn; Werner Bittner at the Deutsche Lufthansa Corporate Archives in Cologne; Mary Herbert at the Maryland Historical Society in Baltimore; David Spencer and Larry Wilson at the National Air and Space Museum in Washington, D.C.; Frau Magister Keller and Frau Winkelbauer at the Austrian State Archives in Vienna; Frau Dr. Keipert and her staff at the German Foreign Ministry Archives in Bonn; Dr. Bourgeois at the Swiss Federal Archives in Bern; Dr. Glaus and his staff at the Federal Institute of Technology Archive in Zurich; the staff of the Archives of the French Air Force, Paris, and Madame de Ruffray, of the Oral History Section; Larry D. Sall and his staff at the University of Texas at Dallas History of Aviation Collection; Frau Dr. Stöckl at the Technology Museum in Vienna; Dr. Meighörner, director of the Zeppelin Museum in Friedrichshafen; Dr. Blübaum and Frau Magister Waibel at the Zeppelin Archive in Friedrichshafen; and Leonard C. Bruno and the staff of the Manuscripts Division at the Library of Congress, in Washington, D.C. The Stinnes family kindly granted me access to the Hugo Stinnes papers held at the Konrad Adenauer Foundation in Bonn.

Several institutions provided financial support for research. My gratitude goes to Helmuth Trischler, director of the Science and Technology Research Institute at the Deutsches Museum, who generously funded a stay in Munich in 1994. The National Air and Space Museum of the Smithsonian Institution provided funding for an archival visit in 1997. The Boston University Engelbourg History Award funded research at the Library of Congress. The Professional Council of Albright College, through a grant of the Christian A. Johnson Foundation, helped support three research and writing trips to Europe in 1996, 1997, and 1998. Hartmut Lehmann, director of the Max Planck Institute for History in Göttingen, offered a summer fellowship toward the completion of the book.

Let me also extend my thanks to the scholars, experts, and enthusiasts who helped along the way through discussions, suggestions, and the sharing of references: Carl J. Bobrow, Peter Capelotti, Tom Crouch, R. E. G. Davies, Hans-Liudger Dienel, Marc Dierikx, John Duggan, Dorothea Haaland, Von Hardesty, A. J. Kox, Cornelia Lüdecke, Paul Maravelas, Karl von Meyenn, Robert C. Mikesh, John H. Morrow, Günther Ott, Dominick Pisano, Douglas Robinson, Frederick M. Schweitzer, C. G. Sweeting, and William F. Trimble.

Peter Fritzsche provided important comments at a critical juncture in the writing of the manuscript. Alf Lüdtke made helpful and welcome suggestions on refining my analysis. Henry Cord Meyer shared his insights during a multiyear correspondence regarding airship politics. Peter and Lilian Grosz welcomed me at their Princeton home on very short notice and made my research in Peter's archives most pleasant and enlightening. John Provan opened his remarkable collection of airship memorabilia and provided interesting anecdotes about the Zeppelin era. Arwen Mohun offered me an opportunity to test some of my ideas at a History Workshop at the University of Delaware.

I am especially grateful to Michael J. Neufeld, who made it possible for me to discuss the project at a "works-in-progress" meeting at the National Air and Space Museum. Michael also offered many helpful comments, suggestions, and criticisms throughout the project's duration. My editor at the Johns Hopkins University Press, Bob Brugger, worked with me above and beyond the call of duty. He suggested turning the dissertation into a book and patiently granted me numerous extensions while helping to improve many sections of the manuscript. Mary Yates also contributed heroically to the completion of the project, not least through her careful editing of my Gallicisms.

I am also indebted to a wide circle of friends and family who provided me with critical comments, bibliographic references, and encouragement throughout the course of my research and writing: Misty Bastian, Dirk Bönker, Alexei Cowett, Bernhard Debatin, James Dutton, Winfried Fischer, Jussi Hanhimäki, Paul Jaskot, Peter Kirchgraber, Fred McKitrick, Carolyn and Paul Mitchell, Guy and Jacqueline Princivalle, Stefan Rinke, Holli Schauber, Greg and Maia Schroeder, Kathryn Slanski, Patricia Stokes, John Svatek, Katharina Tumpek-Kjellmark, Aleksandar-Saša Vuletić, and colleagues and friends at the Einstein Papers Project in Boston and at Albright College. Finally, Maria Mitchell provided love, companionship, and a healthy dose of criticism and encouragement, even going so far as to develop an interest in the history of Zeppelins. For that and for being there from the beginning until the end, I thank her with all my heart.

Through their love of books and *anciennes choses,* including some of the pictures that illustrate this study, my family instilled in me the love for history that first brought me to this project (although they are still trying to figure out where the aviation part came from). Cousins Brigitte and Philippe Sion forwarded me information gleaned through their own studies; my brother Edouard provided the musical entertainment, while Richard Connolly prepared award-winning dinners for my visits. My mother, Joëlle de Syon, discovered and set aside many rare documents I would not have found otherwise. Through her support I was able to complete my education in the United States and follow in her path of tracking down *la petite histoire.*

ZEPPELIN!

Visions of the Sublime

ARLY ON 18 SEPTEMBER 1928, NEAR THE southern German town of Friedrichshafen, several hundred men gently pulled a giant dirigible, the brand-new LZ 127, from its shed. Christened *Graf Zeppelin* in memory of Count Ferdinand von Zeppelin, founder of the firm that had built the machine, the airship was about to undertake a thirty-six-hour test flight from the Rhineland to Hamburg, Berlin, and back before embarking on a crossing of the Atlantic that would demonstrate to the world the value of airships in modern, long-distance air transportation.

For the head of the company, Hugo Eckener, the first, shorter journey had an important public-relations purpose. The company could not have built the new machine without assistance from German taxpayers and private fundraising efforts. "Millions of shareholders," said Eckener, clamored for flights over their own cities; flying the dirigible over Germany would acknowledge and reward their support. Once aloft on its lazy course over the heart of the Father-

A cloud shaped like a reclining Count Zeppelin looks down upon his namesake airship in this 1928 illustration from the magazine *Simplicissimus*. The illustration was published on the occasion of the new LZ 127's trial flight around Germany in September of that year. ASTRA land, the machine did in fact draw thousands of ecstatic onlookers. They filled streets below, waved flags, and shouted happily at the sight of what seemed to be a stupendous example of peaceful German nationalism. The German magazine *Simplicissimus* illustrated the event by depicting a cloud shaped like the departed Count Zeppelin looking down approvingly on the new machine. Twenty years before, the count's fourth dirigible—despite crashing on the return leg of an endurance flight—had launched a popular outpouring of Zeppelin fever that lasted throughout the imperial era. Now, nearly a decade after the Treaty of Versailles had put an end to most German air activity, the "giddiness" was back, Eckener noted happily, "and stronger than ever!"[1]

From 1900, when Count Zeppelin first flew a rigid airship, until many years afterward, aviation progressed along two separate paths: that of the airplane and that of the airship. At first, the airship, which had developed steadily since the late nineteenth century, appeared the more promising machine. By the onset of World War II, however, the airplane had overtaken the dirigible by almost every measure, bringing the golden age of airships to an end. After the war, lighter-than-air flight meant puttering about in blimps, small, nonrigid craft that served mundane purposes like coastal patrolling and advertising at sporting events.

Accounts of the mighty Zeppelin, if fairly abundant, typically supply a mix of adventure story and technical catalogue, reprocessing the information contained in contemporary accounts while oddly avoiding the most interesting question of all: Why did these machines exercise such a powerful influence on the popular imagination?

As this study makes clear, the rigid airship made one of the strongest impressions on European collective memory of any machine, especially in pre–World War II Germany, where the Zeppelin and its inventor enjoyed extraordinary levels of popularity. Only 119 Zeppelin rigid airships were built between 1900 and 1939, yet Germans celebrated them in countless kitsch pieces—mass-produced collectibles that, just like battlefield souvenirs or mementos of Queen Victoria's golden jubilee, illustrated their subjects' hold on the popular mind. While relatively few people flew in the "flying cigars," the masses adopted them as their own. Zeppelins may have transported thousands, but they enchanted millions. The airship's symbolic functions exceeded its practical use; like bicycles, household tools, or other objects of everyday life, its influence remained unmeasurable.[2]

The Zeppelin was too important not to appear in public discussion, but it meant different things to different people. Pacifists believed that airships eventually would effect a new sense of interdependence and produce world unity. Some militarists hoped that the airship would become Germany's ultimate weapon. Shrewd businessmen saw it as a flying piggy-bank, while dedicated technicians of buoyant flight defended its potential against that of fixed-wing aircraft. All commentators agreed on one thing: Count Zeppelin and his invention offered the spectator a vision of the technological that transfixed the imagination.[3]

Recent attention to the sometimes "sublime effect" of technology owes much to the work of an American historian of technology who in the mid-1990s observed a mix of popular fascination with and fear of great-sized projects.[4] "Bigness" boggles the imagination: it requires us to adjust our powers of perception to an object so huge that it leads to a state of near rapture.[5] Examples of technological gigantism abound in North America, from the Brooklyn Bridge to rocket launches. Most Americans do not view such things as invasions of their environment; rather, they have adopted them, walked on them, witnessed their operation on television.[6] In the technological sublime, nature and civilization coexist; they also become part of social identity. Yet Americans have no exclusive claim to the phenomenon. Imbuing technology with patriotic and democratic values, viewing it as a force directing the nation's destiny, also characterized late-nineteenth-century Europe. The industrialization of European cities, for example, while reflecting control of the nation-state, also acquired a popular dimension, whereby the masses assimilated certain human constructions, from train stations to bridges, as symbols of their communities.[7] In Germany, the Zeppelin became the icon of choice. The "Zeppelin spirit" grew out of the majesty, the incredible size, of the machine, along with its technological sophistication. Before World War I and afterward, it permeated the awareness of a German public eager to celebrate an engineering marvel that seemed "all German."[8]

How and why Count Zeppelin and his machines became symbols of national progress also sheds light on the dynamics of German society under different regimes.[9] Between 1900 and 1939, Germany's government was, by turn, authoritarian, democratic, and fascist. In all three systems the Zeppelin's grip on the popular imagination remained strong, and yet Wilhelminian, Weimar, and Nazi Germany provide dramatically different sociopolitical and cultural contexts for examining the evolution of a new technology and its societal impact.

The technological sublime as an element in popular culture may belong to modern politics generally; in this case, it serves to reflect competing definitions of Germanness. In a young state struggling to define its boundaries and identity, symbols abound—and machines may even acquire meanings that do not match the visions of a government. At any rate, Germany in the first half of the twentieth century offers an especially valuable case study of the intersection of technology and culture.

Exploring the interwoven fabric of aviation technology and society,[10] this book focuses on the Zeppelin spirit and the forces that shaped it.

CHAPTER I

Balloons into Dirigibles

D

URING THE LIFETIME OF COUNT FERDINAND von Zeppelin, 1838 to 1917, Europe industrialized and new means of transportation appeared, first the railroad, later the bicycle and the automobile. Speed was their main advantage. Travel from Berlin to Munich would eventually take hours, not days, and other inventions such as the telephone and electric lighting further broadened the horizons of well-to-do Europeans: the dream of travel and exploration became a reality through new technologies.[1]

Count Zeppelin experienced these processes. One of the most dramatic developments of the industrial age—controlled flight—did not occur until he reached retirement age. Yet the count and his contemporaries were exposed throughout their adult lives to the notion of flying contraptions, both in print and in reality. This setting—this mix of invention and fantasy—influenced public views of aeronautics and paved the way for Zeppelin's own work and hard-won success.

The Floating State

Human flight in its first incarnation, the balloon, was achieved during the Enlightenment, but other means of flight did not come into existence until much later. This gestation period reflected a peculiar intersection of popular interest and advanced science, of mechanized technology and scientific endeavor. What would-be pioneers of flight saw as a great opportunity to offer humankind new means of locomotion and exploration, public culture first perceived as entertainment.[2] In fact, the slow, indirect extrapolation of dirigibles from ballooning throughout the nineteenth century established the basis for the acceptance of Zeppelins. The process was both technical and cultural.

Histories of aviation often treat lighter-than-air flight as a side road on the way to routine flight, thereby overlooking the challenges surrounding the resolution of early aviation problems.[3] By the nineteenth century, scientific understanding of flight had made great strides, but transference of this knowledge into practical testing failed. Hoping to succeed through trial and error, practitioners of applied aerodynamics experimented without understanding the principles of their field, namely, fluid dynamics.[4] A similar approach affected the realm of lighter-than-air flight.

Of course, the endeavor appeared simple at first. When, in 1783, the Montgolfier brothers built the first balloon capable of sustaining humans in flight, it suggested to many Enlightenment-inspired observers that humanity had triumphed over the last great frontier. In both Europe and North America, "aeronauts" took to flying for prestige and entertainment. Once airborne, however, they all encountered the same challenge: wind. One could pick at will a place to lift off in a balloon, but there was no guarantee of landing near any hoped-for destination. Controlled directional flight did not exist.

The challenge of designing an effective "powered balloon" capable of withstanding heavy winds came to consist of two related issues: navigation and altitude control. Where navigation was concerned, shape mattered. Based on the principles of teleology, the philosophical discipline that searches for evidence of design in nature, several flight pioneers suggested an ellipsoid balloon that might "swim" in the air like a fish in water. Jean Baptiste Meusnier proposed such a project in 1783, and two decades later, long before wind tunnels first appeared in the late nineteenth century, the Englishman George Cayley attempted to study optimum shapes of least resistance.[5]

Meusnier also helped lay a foundation for proper altitude control. Early means of height adjustment were very inefficient. To gain altitude, the balloon's occupants threw out ballast. If they dumped too much, they became trapped in the air, having no means either to recover the lost ballast or to vent hydrogen from the envelope. Meusnier's solution involved trapping hydrogen in an internal envelope, called a ballonet (gasbag), surrounded by air sealed in the outer envelope. Air pressure could thus be increased for descent or decreased for ascent without sacrificing precious hydrogen. Meusnier's plans were never put into practice—he was killed during the French revolutionary wars—and balloonists adopted the simpler principle of a mechanical valve at the top of a balloon that released hydrogen on command. Yet Meusnier's ballonet solution, along with the ellipsoid shape he suggested, would become central to the design of dirigibles.[6]

Seeking a solution to the problem of flight eventually came to form one example of science-based technology, which appeared in the nineteenth century: to be able to fly implied an understanding of basic scientific principles combined with mechanical skills. Throughout Europe, learned societies spent time discussing the possibilities of controlled flight, whether using balloons or heavier-than-air constructions. An "aviator" at the time was either a scientist who used ballooning to study the atmosphere or a natural historian who studied the flight of birds and determined a pattern that might be duplicated.[7] Such activities were common in many nations, but outsiders found it difficult to take seriously the activities of enthusiasts who looked more like avid bird watchers than would-be masters of the skies.[8]

The problem was exacerbated in the case of ballooning. The aeronautical field in the nineteenth century was a hodgepodge of hypotheses, fantasist conceptions, and dissertations about controlled ballooning. These ranged from Ernest Pétin's balloon train, a multisphere monstrosity that drew considerable public support, to Joseph Kaiserer's proposal to train eagles to pull a balloon carriage-style.[9] These and other ideas reflected a kind of "balloonacy" that turned nineteenth-century aeronautics into an eccentric's field: registered flying patents offered little if any improvement over previous suggestions for solving the problem of dirigible flight.[10]

Propulsion technology reflected such trouble as well. "The dirigible airship is on the whole a utopia!" claimed German skeptics in the late nineteenth century when commenting on the various attempts at steering balloons through the air.[11] For all the imaginative designs in existence, no one had solved the problem of effective propulsion. Many drawings, even those of traditional bal-

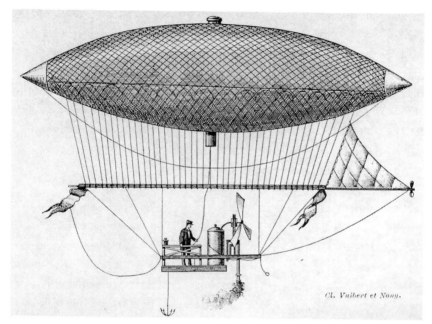

The Giffard airship of 1852. Giffard's was one of the first attempts at streamlining a balloon in order to make it steerable. This picture shows the size of the steam engine, with its chimney turned downward to protect the envelope. ASTRA

loons, included a propelling screw, but would-be inventors overlooked the matter of what, or who, would make the airscrew turn.

Until the invention of the steam engine, alternatives for propulsion of balloon projects usually revolved around some form of human energy. Some designs, based on hydraulic principles, called for paddle wheels, despite inconclusive test results. Other proposals suggested the propeller principle, whereby the balloon crew would rotate a giant screw-shaped cylinder in the manner of a treadmill, with passengers called on to help during ascent and descent.[12] Far more rational than many of his contemporaries, George Cayley considered using a steam engine, but his own calculations suggested that weight made the idea impractical.

The increased understanding of the motor principle in the nineteenth century did not resolve the problem of air navigation and propulsion. Even though the engine had found successful applications in sea and ground transportation, seven decades after the Montgolfier brothers' flight there was still no viable means of air propulsion.[13]

The first successful testing of a practical powered balloon, in 1852 in France, highlighted the propulsion problems more than it solved them. The Frenchman Henry Giffard, inspired by the works of other pioneers, had set about patenting and building an engine-propelled airship. He tested a steam engine placed in a hanging nacelle well below the hydrogen envelope (the engine's chimney had also been inverted to avoid the risk of igniting the hydrogen). Giffard's airship flew several kilometers but failed to make headway against the wind; his later experiments remained inconclusive.[14] So disappointing was the performance of the steam engine that in at least one later case, that of the Frenchman Dupuy de Lôme, the project called for the use of a manually steered four-bladed propeller.

The few innovative ideas that did develop failed in the long run for lack of funding. One example was the designs for "steerable" balloons of the German-born pioneer Paul Hänlein, which represented substantial progress toward an effective solution. Building on the work of Meusnier, Hänlein first tested two dirigible scale models in Mainz and Vienna before initiating full-scale construction in 1872. In the nacelle he placed a Lenoir engine that drove two four-bladed propellers. This radical departure from the steam engine had a built-in propellant source: it used gas drawn from within the envelope.[15] While this principle would later reemerge in a Zeppelin airship design, it failed to convince Hänlein's backers, who accused him of working too slowly and withdrew funding.[16]

Other ideas included the use of an electrical engine. In 1882–83 the Tissandier brothers built and tested a dynamo-powered airship. The device ran up to two and a half hours, but it weighed over 360 kilos and furnished only 1.5 horsepower, which translated into an approximate speed of 15 kilometers per hour (kph). A light summer wind could thus pin the machine down on the spot, and the lack of power prevented further progress.[17] The Tissandiers' failure, however, offered other pioneers food for thought.

In 1884 the Frenchmen Charles Renard and Arthur Krebs devised the dirigible *La France* and installed a lighter (95-kilo), more efficient electrical motor, developing up to 9 horsepower. This allowed for the completion of several successful flights with speeds of up to 23 kph. In addition, Krebs and Renard had taken great care to streamline their design in such a way that the rudder would be more effective. This did not mean that they achieved efficiency; most of the lifting power of the gas was applied to the engine, preventing transport of any passengers but the crew. Renard and Krebs appealed for further funding but received little, suggesting into how much disfavor dirigibility research had fallen.[18] The same fate met Hermann Ganswindt's work on dirigibles, with the German army rejecting petition after petition for funding to support construction of a machine.[19] German authorities remained skeptical in part because of the suggestion of the influential physicist Hermann von Helmholtz that powered-flight experiments were mere stunts with no practical future.[20]

News of these flight experiments, theoretical as well as practical, reached the public through both rumor and media. But reactions to flight in the nineteenth century were an ambivalent mix of enthusiasm and apathy. The response went in cycles. In the late Enlightenment and throughout the nineteenth century, the balloon served as an important escapist symbol to ideologists of every stripe, from rationalists and revolutionaries to conservatives and romantics.[21] To escape gravity meant mastery of nature to one, defiance of law to another, an opportunity to observe the immobile Earth to a third, a romantic experience to a fourth.[22] Ideologies that would normally collide on the ground seemed united in their enthusiasm for flight. This factor of unity would also play a role decades later in Zeppelin's success. Some scoffed at the enthusiasm and noted that flight practitioners did it only for flight's sake, without considering practical applications for their art. As Nietzsche remarked skeptically decades later, one lacks direction when floating in a balloon; the "powerful craving" for flight is nothing but hot air.[23] As for the dirigible, critics suggested that its uses were even less promising than the balloon's.[24] Both might serve for scientific weather investigations, but neither could offer efficient travel through the air.

Popular attitudes began a slow shift in the late nineteenth century, and no longer depended on personal experience—on witnessing a balloon flight or a failed attempt. Throughout Europe, popular literature began to suggest that flying machines would be the next stage of the industrial era. The imaginary airship early on acquired a special place in mass culture, contributing to a sense of the "technological sublime," a perception combining fear of and fascination with a machine, all the while endowing it with almost sentient features.[25] This peculiar imaginative relationship to a class of inanimate objects began before the Zeppelin era but may have contributed to the machine's later success.

Progress through flight was but another expression of machines offering humanity a greater future. The imaginary roles assigned to the airship did not simply denote utopian dreams; they also pointed to a popular need to demystify novelty and place it within parameters understandable to all. This rendering of a fantasy based in reality says much about how Europeans made sense of the new technological age. Indeed—and in contrast to rocket aircraft, which in the early twentieth century came much closer to the norm of full-fledged utopian representations of the future—the airship was a means to utopian imagery rather than an end in itself.[26] Because it suggested that machine dreams might come true, the airship provided a psychological link between fantasy and reality, leading contemporaries to yearn for greater benefit from technological achievements. In this sense, airships, both real and imagined, reflected the atmosphere of progress that pervaded the second half of the nineteenth century.[27]

Not all interpreters of future machinery were optimistic, however. Visions of flying machines—whether in novels or on the collectible cards that cigarette and chocolate manufacturers offered—depicted technology in a variety of ways, from frankly humorous representations to more cautious pictures of a dark future.[28] Many futurists speculated that warfare would soon include air battles, and they did not hesitate to depict airships attacking each other. The increased place of war in futuristic depictions was due in part to the Franco-Prussian War of 1870–71. Prussia's swift dispatch of French troops had shown that a military force relying on new methods of organization and new technologies could gain the upper hand. Albert Robida's designs were among the most famous representations of a future "air age." Reprinted throughout the European press, these projections ranged from cloud hotels to fierce-looking "Hawk" attack dirigibles, all reflecting a peculiar mix of satire and patriotism. Although attractive in their artistic quality, these images also reflected the feeling that machines would soon control the destiny of men: symbols of progress would also signal the decline of humanity.[29]

Popular authors throughout Europe and the United States focused on future wars and populist imperialism as themes of entertainment.[30] In these works of imagination, limited knowledge of the real potential (and drawbacks) of airships led to some odd representations of airships engaged in infantry-like battles alongside standard aerial reconnaissance.[31] Although discounted as generic mass entertainment, such stories and their diffusion through popular-press serialization no doubt instilled in the public a speculative awareness of things to come. Echoing Robida's approach, novelists fascinated readers with a mixture of bright and bleak visions of the aerial future. Edgar Allan Poe, drawing inspiration from the studies of the British air pioneers George Cayley and William Samuel Henson, fooled audiences with his great transatlantic balloon hoax, suggesting in serialized articles that preparations for an Atlantic crossing were under way and recounting the flight itself to a readership eager for sensation.[32] Jules Verne used futuristic airship technology to propel great exploration adventures, such as *The Clipper of the Clouds*.[33] The publication of more serious pieces by respected men like Hiram Maxim, of machine-gun fame, anchored the contradictory notion that the airship would indeed not only fill the skies of the future in peacetime but become an engine of war.

The idea of using flight for military purposes went back to the French Revolution. At the battle of Fleurus, for example, the French army had used a balloon to observe enemy movement. By the 1860s, this experience and other episodes in later wars had inspired the creation of military ballooning units throughout Europe. While bombing was foremost in the minds of strategists, observation was the primary purpose of the balloon battalions. Naturally, military planners had discussed countermeasures designed to bring down any flying object.[34] Partly in response to the bombing threat, the 1899 Hague Convention on the laws of war banned the throwing of projectiles from balloons; to observe enemy troops was one thing, but to hit ground targets from above was deemed dishonorable and unfair. Drafters of the agreement assumed that heavier-than-air machines would not operate in the near future. Indeed, until the Wright brothers demonstrated their prowess in Europe in 1908, most discussants of air warfare thought it would depend on the use of balloons or airships.[35] Such machines would help reaffirm imperialist goals and deter any challenges to authority over new territory.[36]

Ironically, then, turn-of-the-century visions of the airship as a war machine also included an element of idealism that would confirm a country's technical and military superiority. Some optimists even claimed that the very notion of an air threat would help deter any future conflict.[37] Although public interest in airship flight included visions of gloom and doom, the optimistic note dominated, thanks to the fascination with flight itself. To challenge gravity successfully would turn nations into neighbors rather than enemies, for traditional borders would lose relevance. To most observers, only mature industrialized civilizations could purport to take on such a challenge successfully.[38] Progress thus remained the primary notion associated with the flying machine, but it was not the only one.

After the midcentury phase of "balloonacy" had ended, the public continued to view ballooning as a great source of entertainment.[39] Tethered giant balloons became favorite attractions at industrial exhibitions, while aeronauts topped one another in daring balloon stunts, covered in the media as circus events.[40] If the spectators developed a passion for flight, they could easily join

Future visions: the balloon world in 2000. Food companies at the turn of the century offered collector cards as a means of marketing their products. The Swiss brand Sprüngli, among others, focused on futuristic representations. Here, people sail through the air as if taking a leisurely Sunday boat ride. In this fantasy, the future is essentially like the present; aside from the vehicles, nothing has changed, not even clothing fashions. Such illustrations were legion and accustomed consumers to the notion of an impending age of airships. Author's collection

aeronautical clubs. Among the public associations that appeared during the late nineteenth century, aeronautical societies were so common that most major cities throughout Europe could claim either a major association or a chapter. Most who joined, however, claimed a scientific interest in ballooning, and the activities of aeronautical clubs generally focused on scientific measurements of the atmosphere.[41] Societies occasionally planned, but rarely carried out, long-distance flights. When these occurred, they rarely exceeded 300 kilometers. However, as a form of adventure, accounts of such flights riveted audiences and newspaper readers.[42]

Members of flying societies also presented possible solutions to the problem of controlled flight. Enthusiastic engineers considered ways of improving balloon designs, including, of course, means to make them steerable. But the indiscriminate publication of every new theory in aeronautical journals made it difficult for readers to distinguish serious studies from eccentric musings. Nonetheless, through such associations, and through the endeavors, however unsuccessful, of the pioneers—Germans like Gustav Koch and Friedrich Wölfert, or even "mad" King Ludwig II of Bavaria, who sponsored the creation of the Bavarian Airship Flying School[43]—people became fairly well accustomed to the idea of flight.

Even so, by the turn of the century membership in such groups had risen to only a few hundred committed enthusiasts, one of whom was Count Zeppelin.[44] In this broader context of mixed popular and scientific interest, the count's tortuous path to success becomes easier to explain. Whereas idealized biographies paint him as a misunderstood genius struggling against adversity, the real sequence of events simply followed the ebb and flow of public and governmental interest in flying machines during the late nineteenth and early twentieth centuries.

From Enthusiasm to Obsession: The First Zeppelin

Count Zeppelin's aeronautical journey began in August 1863, when, as a young army officer from the southern German state of Württemberg, he traveled to the United States, hoping to observe maneuvers in the Civil War. He received permission from President Lincoln to accompany Union troops and act as a neutral observer. In this capacity, he participated in the launching of a tethered balloon.[45] His first awareness of flying machines and the challenge of steering them probably dates to this event, but he paid no attention to the matter for another decade. In the meantime, he served with distinction in the Franco-Prussian War and, despite misgivings, embraced the unification of Germany under Prussian leadership in 1871.

In 1874 Count Zeppelin recorded his first thoughts on the use of airships. That year, the German postmaster general, Heinrich von Stephan, one of the founders of the Universal Postal Union (and one of the fathers of the postcard), addressed the Berlin Science Society on the theme of "World Mail and Air Travel." In his speech, von Stephan argued that since "traffic and culture go hand in hand, as blood goes through the human brain," mail should take advantage of new technologies to benefit humanity. Reminding his audience that Paris under Prussian siege in 1870 had successfully dispatched over sixty balloons carrying mail and passengers, von Stephan argued the benefits of rapid postal service by balloon.[46]

Count Zeppelin read the postmaster's remarks and became fascinated with the potential for a successful dirigible. While pursuing his military career, he began writing down his ideas about flight in a diary, remarking that his projected vehicle would have to be as "big as a ship."[47] In 1887, concerned at the progress the French were making in aeronautics (the dirigible *La France* had flown successfully three years earlier), Zeppelin wrote a letter to the king of Württemberg, stating the need for a machine that could give Germany the advantage of flying militarily and commercially.[48]

Power politics, however, hurt Zeppelin's career and his ability to disseminate his ideas. Concerned about undue "Prussianization" of the Reich military structure at the expense of local German armies, the count alienated several high-ranking members of the imperial general staff. This rancor came back to haunt him in 1890, when he received a failing grade on his performance in joint imperial maneuvers. Forced into retirement at age fifty-two and unable to make suggestions about flight through military channels, he began to devote himself to a dirigible project.[49] Another decade would elapse before his first machine ascended, a decade during which Zeppelin worked to convince others of the soundness of his views on flight technology.

Histories of Count Zeppelin's dealings with German engineers have generally fallen prey to the myth of the lone genius battling obtuse bureaucracies. Such accounts were based in part on stories the count's associates wrote to commemorate his later success.[50] While bureaucratic challenges were par for the course, such accounts ignore the circumstances: Count Zeppelin was but another inventor with no technical training, one of hundreds claiming to have solved the problem of dirigibility.

The technical challenges Count Zeppelin faced shaped his relationship to the German engineering profession. To obtain state funding, he required the recommendation of a government-appointed engineering commission. His first attempt to obtain approval, made in the summer of 1894, failed. Zeppelin had based his ideas in part on the flights of the *La France* nonrigid airship a decade earlier but had gone one step further, suggesting not just an air vehicle but an air train. One section of the airship, containing the engines and propeller, would tow the other, not unlike a locomotive towing a caboose.[51] Experts immediately pointed out problems with the sheer size of the count's proposed airship and its rigid structure. The count's huge design was to rival *La France* for speed, but in no other way could it match its predecessor's capabilities. Furthermore, since the Zeppelin was intended for military reconnaissance, the 32-kph speed would have still made it an ideal target for enemy fire. In addition, the engineer entrusted with the calculations for the count's project had accounted for vertical rigidity (ensuring that the machine would not bend during altitude changes) but not horizontal rigidity (turns). Finally, there were no provisions for an airship hangar; all previous dirigibles had required some form of protection from the elements while on the ground.

Undaunted, Count Zeppelin took the commission's negative report in stride and submitted an improved design in September 1894. Refined through the inclusion of an aerodynamically shaped nose cone, this design, for a 134-meter-long machine, proposed to use two engines furnishing 9 horsepower each.[52] These engines were to be of the internal-combustion type, a new design, deemed more efficient, that had also been used in other experimental dirigibles.[53]

Through most of the century, steam engines had dominated the mechanical scene. They were massive, requiring not just combustible material (wood or coal) but quantities of water housed in a reservoir and then transferred into a heavy boiler unit. The power-weight ratio, acceptable for a locomotive, was impractical for an airship. Jean Joseph Étienne Lenoir's gas engine, first patented in 1860, consumed far too much oil and grease to be practical either, but it paved the way for improvements through the work of Nikolaus Otto, Rudolf Diesel, Gottlieb Daimler, and Wilhelm Maybach. This work eventually led to the creation of the four-stroke cycle engine (wherein cylinders transform the fuel into propelling energy during the third, internal stroke).[54] When inventors began switching to gasoline as a fuel source, the internal-combustion engine became a promising means of efficient propulsion. Despite low performance initially, it would soon compete with and displace the electrical engine, and usher in the transport revolution. Relying on the work of Daimler and Maybach, Carl Benz patented his first automobile in 1886 and founded a car factory in 1890, naming his product after his daughter, Mercedes. The maximum speeds attained by the Daimler engine barely reached 20 kph on land, but it proved so much more efficient than the steam engine that Zeppelin hoped to use it on his invention.

Zeppelin staked his hopes on this promising innovation, but his 1894 project failed to account for mass. The average weight of the combustion engines required to power such an airship was 500 kilos each, not counting fuel and water for the radiators; there would barely be enough capacity for the weight of even a small crew. The commission immediately noticed the problem and asked whether model testing might first take place to work it out, but the count wanted a full-size prototype. Finally, although several of its members were sympathetic, the commission declined to recommend sponsorship of the project, and the Prussian War Ministry followed suit.[55]

Several more exchanges between the count and military authorities failed to resolve the matter in his favor; the idea was simply too fantasist. One officer even scribbled the comment "Jules Verne" in reference to the numbers supplied.[56] Engine power and weight had brought the count's idea down, yet he chose to file for a patent, granted in 1895. Years later, when analyzing the count's subsequent success, commentators would focus on the expert commission's judgment and attack its claims that Zeppelin's plan was impossible. In fact, the commission was entirely justified in its conclusions. The project was unrealistic in its existing form, and Zeppelin would have to modify it several times.

Undaunted by the experts' negative appraisal, the count sought and obtained the cautious support of the German Engineers' Association (VDI). As one of two major technical associations in imperial Germany, the VDI might turn the tide in Zeppelin's favor. Indeed, the influence of pressure groups was an important force in German imperial politics. Since many VDI members' specialization bore no relationship to flight technology, all Zeppelin needed was qualified approval of his proposal to claim the support of the entire group. He could in turn use this support to counter the expert commission's results.

"An airship, to be useful, must be able to lift off into the air," quipped the count to an audience of over four hundred engineers in February 1896.[57] The joke reflected the general state of aviation, whereby hundreds of would-be fliers had failed to become airborne, losing fortunes and becoming the butt of endless ridicule. Moving on from humor, the count explained that the solution should allow vertical as well as horizontal movement. His design had undergone further revisions and now consisted of a rigid internal structure built of seamless tubes, wire ropes, and flat rings. Further subdivisions into cylindrical chambers enclosed the hydrogen gasbags and gave the structure additional rigidity. The cylinder was covered in fabric and had two payload cars affixed underneath, each intended to carry between 600 and 2,000 kilos. Two propellers linked to one engine (to save weight) would provide thrust.

Although this project was based on the 1895 patent, the count acknowledged that he had relied on other patents to devise it.[58] That no dispute over patent rights ensued has a simple explanation: Any would-be inventor could file a patent for a fee. It meant little so long as he did not build the patented device. It also meant that professional engineers might not see any more validity in Zeppelin's project than in any of the other flying-machine patents collecting dust on shelves; that a device was patented was no guarantee of its practicality.

Nonetheless, the VDI's response to the count's presentation was cautiously positive. The report limited itself to the merits of Count Zeppelin's proposal and steered away from any criticism of his technical qualifications. Stressing the need for greater involvement from industrial and engineering circles, the report's closing sentences presented the matter in a new light: "France, North America, and England have overtaken us [in the field of flight] with considerable resources. Should not German technology participate too in the solution to this problem?"[59] Although strong patriotic support for the count's efforts was still far off, the matter had already become a question of honor, of Germany's duty to respond, almost in the fashion of a duelist, to the challenge of sustained flight.

Other Germans had become interested in the matter and offered alternative solutions to Zeppelin's. In 1897, for example, Friedrich Wölfert and his assistant Robert Knabbe tested an airship propelled by a petrol-fueled internal combustion engine. The heat of the device, however, ignited the hydrogen in the gasbag, and both men were killed in the ensuing explosion.[60] That same year, another solution was tested, one that involved an aluminum-clad airship. The idea of building an airship envelope out of metal dated back to the late Enlightenment.[61] It was not until the appearance of aluminum in the late nineteenth century that inventors started considering the twofold advantages of a light metal casing. First, it would ensure that the dirigible envelope would keep its shape regardless of the gas pressure inside the hull, thus preventing steering accidents. Second, such solidity would also keep the hydrogen contained within the envelope—or so the theory went—until released through a trap for descent, thereby reducing the risks of a sudden fall. Regular gasbags were not fully airproof, as they were made of goldbeater's skin, taken from cattle bladders. The material was impermeable but fragile and could suffer tears during manufacture and pressure filling. Furthermore, it was extremely expensive, several thousand skins being required to produce a single gasbag (the use of treated silk would not come till much later). Hydrogen containers of aluminum thus might have resolved a serious problem.

The aluminum-clad airship was the brainchild of the Austro-Hungarian-born inventor David Schwarz. Active in the wood-manufacturing business, Schwarz developed an interest in dirigible flight after reading technical publications that discussed aluminum. He offered his project to the Austro-Hungarian War Ministry, which turned him down, but he then caught the attention of the Russian military attaché in Vienna. With the latter's support, Schwarz moved to St. Petersburg and obtained funding for the construction of a dirigible. Completed in 1894, the design leaked its hydrogen furiously and never flew.[62] Schwarz fled Russia and made his way to Berlin, where he

David Schwarz's aluminum-clad airship. Schwarz's first design never flew. After his death, his widow worked with the industrialist Carl Berg to sponsor the testing of a second machine in Berlin in 1897. The artist has depicted the buckling of the aluminum sheets prior to the launch of the dirigible. A combination of leaks, heavy winds, and pilot inexperience caused the destruction of the machine and the abandonment of further development. ASTRA

arranged construction of a second airship. He died in 1897 during a trip to Vienna, before the machine was finished. Soon after, his widow oversaw the completion of the project, with the assistance of the German industrialist Carl Berg. The airship's hull structure was made of aluminum, as were several sections of the Daimler engine and the propeller blades. The hull itself turned out to have leaks due to soldering and buckling problems near stress points.[63] Nonetheless, a test flight took place in Berlin in November 1897. Unfortunately, the machine crashed as a result of strong winds and the inexperience of the military balloonist piloting it. The mishap ended the Schwarz endeavor but did not discredit the use of a light metal in airship construction.

Carl Berg, who had supplied the aluminum to build the Schwarz machine, kept arguing that this new material could indeed aid aeronautics. Aware of Count Zeppelin's efforts, he had offered Zeppelin the opportunity to participate in the Schwarz endeavor. After the project's failure, Berg, eager to market his product, proceeded to supply the count with the metal he needed to make his project a reality.[64]

By 1898, Count Zeppelin had run out of patience and decided to build his machine without government assistance. Taking into account some of the VDI's suggestions for improvements, he recruited technicians to work on the plans and sought investors by establishing a "Society for the Promotion of Aerial Navigation."[65] With enough capital gathered, construction could finally proceed. The king of Württemberg, Count Zeppelin's home state, authorized him to build a floating shed at Mansell on Lake Constance.

In the hangar, LZ 1, as the project was dubbed—a giant cylinder 128 meters long and almost 12 meters in diameter—began to take shape. The contraption consisted of sixteen polygonal rings spaced some 8 meters apart and linked by metal rods. The frame was covered in mesh-net and contained the hydrogen gasbags to an aggregate capacity of 10,000 cubic meters. The load was distributed between two gondolas to ensure stability; each contained an engine that drove two propellers. To maintain proper weight distribution and protect the machine from unintended shifts during climb or descent, technicians strung a mobile weight between the two gondolas. Its controlled movement would allow the crew to point the dirigible up or down, for there were neither elevator surfaces nor an empennage to perform the task. Two tiny vertical rudders at the top and bottom of the fuselage would ensure directional control.

Zeppelin's decision to follow Schwarz and use aluminum may have seemed like a daring choice, but it was in some ways quite logical. Aluminum was still a novelty in the late nineteenth century, a material for which manufacturers were still trying to imagine applications. These ranged from light mechanical structures to jewelry and even postcards. An airship was just one more possible industrial use for the new material.[66] Moreover, although scientists and engineers did not yet fully understand the dynamics of the new metal as they would apply to the rigid internal structure of an airship, the deficiencies of the alternative materials—wood and steel—were well known. Specifically, the vulnerability of wood under humid conditions and the excessive weight of steel made aluminum a better choice. The tensile strength of the metal (its capacity to withstand the application of force over a defined area), along with its stiffness and light weight, made it far more attractive than any other material, despite its novelty.[67] However, the size of the machine Zeppelin had in mind and

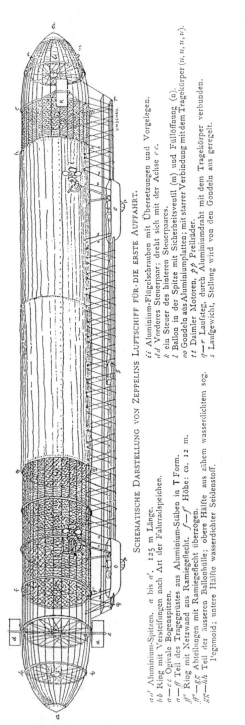

SCHEMATISCHE DARSTELLUNG VON ZEPPELINS LUFTSCHIFF FÜR DIE ERSTE AUFFAHRT.

aa' Aluminium-Spitzen. a bis a'. 125 m Länge.
bb Ring mit Versteifungen nach Art der Fahrradspeichen.
$a — cc$ Ogivale Bogenspitzen.
$n — ff$ Teil des Traggerüstes aus Aluminium-Stäben in T Form.
ff' Ring mit Netzwand aus Ramiegeflecht. $f — f$ Höhe: ca. 12 m.
$ff' — gg$ Abteilungen mit Ramiegeflecht überzogen.
$gg — hh$ Teil der äusseren Ballonhülle; obere Hälfte aus zähem wasserdichten sog. Pegamoid; untere Hälfte wasserdichter Seidenstoff.

ii Aluminium-Flügelschrauben mit Übersetzungen und Vorgelegen.
dd Vorderes Steuerpaar; dreht sich mit der Achse rc.
k ein Steuer des hinteren Steuerpaares.
l Ballon in der Spitze mit Sicherheitsventil (m) und Füllöffnung (n).
oo Gondeln aus Aluminiumplatten; mit starrer Verbindung mit dem Tragekörper (u, u, u, u).
tt Daimler Motoren, pp Prellräder.
$q — r$ Laufsteg, durch Aluminiumdraht mit dem Tragekörper verbunden.
s Laufgewicht, Stellung wird von den Gondeln aus geregelt.

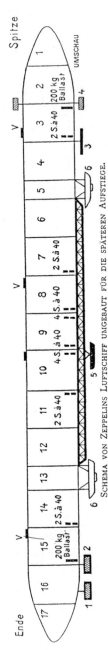

SCHEMA VON ZEPPELINS LUFTSCHIFF UMGEBAUT FÜR DIE SPÄTEREN AUFSTIEGE.

Schematics of LZ 1 for public consumption. These drawings appeared in German (and other) popular-science magazines, yet they contained some artistic fantasies. The upper drawing, for example, misrepresented the moving weight (S) intended to help direct the ship up and down. On the first flight this weight was in fact suspended away from the fuselage. It was not until the second and third flights that a small wagon on rails was added (S on the profile). Despite modifications for the second and third flights (darker lines in the profile), directional stability remained poor. It was not until the construction of LZ 3 that Zeppelin engineers began adopting directional flaps, inspired by French dirigibles. *Die Umschau*, via ASTRA the physical forces affecting its operation meant that use of the new metal would increase the challenge of testing such an airship.

The sheer physical size of the undertaking drew considerable attention over the eighteen-month period separating hangar construction from the first flight of LZ 1. Thanks to the increasing discussion of aeronautics in scientific publications and the popular media, the Zeppelin project was a subject of intense general interest from the very start. Yet the huge floating hangar, resting on ninety-five pontoons, would not yield up its secrets to the curious bystanders who hired rowboats to get a closer look; additional curtains had been hung over all openings. The first flight, initially planned for fall 1899, was pushed for technical reasons to spring, then summer 1900.

To reporters, the fact that such a machine might fly at the dawn of the new century offered new promise that Germany could still catch up in the race to colonize the skies. Mass media rushed to cover the event and repeatedly printed reams of technical facts along with pictures of the hangar, of the motorboat used for testing the airship engine, and of the count himself.[68] The discourses on progress that accompanied such articles excited readers, many of whom sought to witness the airship's first flight.

By 30 June 1900 the banks of Lake Constance, already a favored summer excursion site for the urban elite, were overrun with crowds drawn by rumors of an impending flight. Yet the planned liftoff date came and went because of technical problems and unfavorable wind conditions, causing many onlookers to return home. Capricious weather curtailed attempts the next day as well; a blue flag atop the airship hangar signaled the cancellation of the day's attempt. But by early evening on 2 July, wind conditions had improved. The airship was towed out of its hangar. Around eight o'clock, following the "Los" (let go) command, it slowly took off, rising suddenly to over 300 meters using the full power of its engine to fight a 20-kph wind. Suddenly, one of the engine crankshafts broke, and the machine started to veer suddenly, bending part of the rigid aluminum structure. Nevertheless, the machine had proven that it could be steered in the air. Twenty minutes later, the flight was over.[69] The press, from city tabloids to such staid periodicals as the family-oriented *Gartenlaube*, reported on the event at length: a building-sized machine led by a veteran hero of the Franco-Prussian War had successfully defied gravity![70]

Repairs were carried out and a few technical modifications installed (the hanging weight now moved along a small gangway between the gondolas), and LZ 1 flew twice more in October 1900. By then Count Zeppelin had run out of funds. Although the king of Württemberg and a few military and technical experts believed that the airship would mature if given time, a majority remained skeptical about the machine's ability to defy wind.[71] True, it had not only flown against the wind but had also shown that it could descend and land better than any airship built to date. Yet it was widely felt that these achievements would never be more than a subject of discussion among engineers and enthusiasts.

Several factors explain why a "Zeppelin craze" did not begin at this point. The first reason is summed up in the German word *Gestalt* (mood or character): the inventor and his machine lacked a special identity, a personality that could inspire enthusiasm. That a second flight did not occur for months suggested to some that the experiment was a failure, and supporters of the "Valiant Count" remained a minority throughout the German-speaking world.[72] Although respected as a war hero (his military record was mentioned in schoolbooks),[73] the count had no credentials as a technician. The support he had garnered from the VDI as a way to offset his inexperience was at best sympathetic, but not enthusiastic.

A second reason for the Zeppelin's limited success at this time is that, as it stood in 1900, the Zeppelin project was but the next stage in the evolution of the dirigible. Hundreds of engineers, scientists, and enthusiasts had discussed, patented, and even built lighter-than-air machines, yet they all had failed to turn the airship into a practical flying system. Count Zeppelin was merely joining a crowd. His solution had yet to assume a shape and a dynamic that would distinguish Zeppelins from other airships in the public eye.[74] The transformation of this engineering experiment into a social experience started in 1900, but it would take years to mature as a technological and human symbol.[75]

Third, although the first Zeppelin airship flight coincided with the new century, the cultural climate of Germany had become increasingly infused with a conservative view of technology. Critics viewed industrialization with suspicion, as a force that threatened the beauty of nature and corrupted the simple life that had existed in some supposed golden age. The romantic prejudice against the machine defined it as hostile and depersonalizing to the world of feeling.[76] Mass production destroyed the work of art; the machine could duplicate form, but not sentiment. Technology also challenged the imagined romantic traditions of a young nation. In reaction to this particular form of nostalgia, many Germans sought to rebuild a connection to nature through sports and open-air activities. (This escapist tendency was not unique to Germany; it figured in similar nationalist revivals in France and Switzerland.) Ironically, of course, excursions to the nonindustrial countryside tended to use technologies (railroad and car) derived from industrialization.[77] Yet the mistrust of technology, and thus of technologists, was sufficiently widespread that even a technology as fantastic as flight would not easily enter escapist popular culture.[78]

Finally, the early lack of public enthusiasm for flight in Germany reflected attitudes throughout Europe. All across the Continent, the initial flicker of public interest in flight had given way to disappointment and boredom. Figures acknowledged today as major pioneers, such as Otto Lilienthal and Clément Ader, were at the time mostly known for their failures, and their work attracted but a few thousand believers. By 1903, the year the Wright brothers first flew, the trend would begin to reverse itself, and public interest would grow again. But for now, despite regular media coverage of attempted flights and ongoing speculation about the future of air travel, popular enthusiasm for such endeavors remained low.[79] Moreover, the fact that for years no country gained a clear lead in any kind of aeronautics—France's 1902 Lebaudy airship, although admired, remained a toy of sorts—meant that the German imperial government had no compelling nationalist reason to assist Count Zeppelin. The count's challenge, then, was not just a matter of proper engineering and funding but also one of persuasion.

Following the third and final flight of LZ 1, one frustrated engineer sympathetic to the count wondered whether "the German Michel [the average German] may have to wait again until foreign powers take control of the affair, and then later, at great cost, purchase the business."[80] By 1901, the state of Zeppelin's finances required him to shut down the operation. He let go of most of his engineering help, including his designer, Theodor Kober. A sympathetic Kaiser Wilhelm II awarded Zeppelin the Red Eagle First Class decoration for his efforts—still one level below the more highly regarded Black Eagle order—but no state funding came through.

Building a Zeppelin Culture

In 1902 Count Zeppelin began actively seeking financial support for a new venture in nongovernmental circles. In an address to the German Colonial Society, he described the successes and shortcomings of his first machine. He then argued the need for an airship by invoking the dual ideals of science and nationalism. Germany needed to live up to its earlier technological contributions—train, ship, and automobile—and pioneer the conquest of the skies. Funding the airship would not only make Postmaster von Stephan's dream of airmail a reality; more important, it would give Germany a military advantage.[81] This and other lectures set the pattern for Zeppelin's appeal to both the military and the public for years to come.[82] At the same time, the count also sought to obtain stronger support from the VDI. The revised report he received remained sympathetic but noncommittal, stating that a greater flight speed would prove the capacity of his invention. To the old man, this amounted to a new rejection.[83]

In 1903, frustrated by the skepticism with which his efforts had been greeted in industrial circles, the count made his first attempt to appeal directly to the German public for funding.[84] Reaching the public through the media seemed to offer a solution to his problem. With the help of his lawyer, Ernst Uhland, Zeppelin dealt with the Scherl publishing company (which printed the daily *Der Tag*), obtaining mailing lists and preparing articles on the value of his airship system. However, the fund-raising drive went poorly. Despite Scherl's suggestion that the count's articles should reflect a joyful "heart and soul," their tone was dull and defeatist. Readers of the appeals waded through confusing technical descriptions, only to gather the impression that they were being asked to help an old man who was feeling sorry for himself, not a brilliant inventor who was sacrificing everything for the good of Germany.[85]

Further events strained the count's relationship with his publisher. Publishing mistakes (a misprinted article) and Zeppelin's wish to launch a campaign in August (a vacation month when few read the papers) culminated in a showdown about the publication of further articles on airships. The count, aware that the appearance of his name in the byline would identify him as reporting on his own efforts, wanted Scherl to write the pieces, hoping that this would spark "outsider interest." The publisher refused, and the relationship soured.[86] As for the public reaction, the general response to the mass mailings that accompanied the press campaign was annoyance at being bothered about such useless things.[87] Several respondents expressed the cynical opinion that if God had meant humans to fly, He would have given them wings.

In an attempt to reverse his misfortunes, Count Zeppelin switched to a lottery system to raise funds, but the returns once again were quite meager. As one sympathetic commentator pointed out, such results were typical of reactions to requests for aeronautical funding.[88] However, a shift in technical and military thought during 1903–4 worked in the count's favor. By combining lottery funds with a mortgage on his wife's estate and a few state funds, Zeppelin

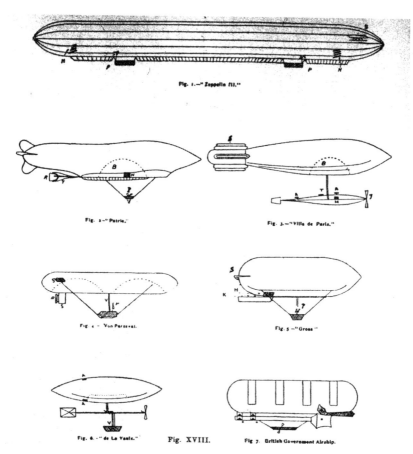

Fig. 1.—"Zeppelin III."

Fig. 2.—"Patrie."

Fig. 3.—"Villa de Paris."

Fig. 4.—"Von Parseval."

Fig. 5.—"Gross"

Fig. 6.—"de La Vaulx."

Fig. XVIII.

Fig. 7. British Government Airship.

Multiple airship solutions. This 1907 scale representation gives a clearer sense of the proportions of a Zeppelin in relation to French, German, and British semi- and nonrigids. Although more practical in field deployments, non-Zeppelin airships had the same weakness as their rigid competitors: vulnerability to wind. More accidents were caused by weather conditions than by any other factor at the time. ASTRA

was able to build a new airship.[89] In so doing, he started another cycle of popular and scientific awareness of the airship.

The government's decision to become involved in Count Zeppelin's LZ 2 project reflected the slow shift in official attitudes toward flight technology. By 1904, three dirigible systems existed: nonrigid, semirigid, and rigid. The nonrigid machines were the earliest design, and the most common. The envelope's shape was maintained through internal gas pressure (the hydrogen itself being contained within ballonets inside the hull). Carrying capacity was limited, how-

Deutsche Einigkeit   E. Wilke

John Bull: „Mit dem derzeitigen Stand der deutschen Luftschiffahrt sein ich sehr zufrieden!"

Three inventors duel from aboard their respective ships. This 1906 caricature satirizes both notions of German unity and the proliferation of airship formulas: Zeppelin's rigid, Major Gross's semirigid, and Parseval's nonrigid. Riding the Prussian eagle, General von Einem adjudicates but seems to point his gun at Zeppelin. Below, the British John Bull expresses his satisfaction with the state of German aeronautics. Thanks to such discord, Albion is a happy island. BA, R5/3829

ever, and there was a risk of hull buckling if pressure dropped, a fairly common occurrence that brought many flights to an uncomfortable end. (Schwarz's 1897 aluminum-clad design promised to remedy this problem.) Yet the nonrigid option was also cheap to build. In Germany, August von Parseval used it in his designs, which reached up to 6,500 cubic meters in volume. Parseval's solution to the problem of hull buckling was to overpressurize the hull so that if a leak occurred, pilots would still have time to reach the ground, even under windy conditions.[90] The formula's simplicity attracted military interest, leading to contracts for a prototype and eventually series production.

The semirigid system attempted to solve the buckling problem by attaching an inflatable hull to a rigid bottom keel, from which the passenger and engine nacelles were suspended. The design was fairly successful, and the French dirigible designers Henri Julliot and Paul Lebaudy sold several in France and abroad, although they turned down a reported German request for a machine.[91] In Germany, Major Hans Gross and engineer Nikolaus Basenach persuaded the army to issue a contract for testing a semirigid solution, as it might offer the best of both worlds: capable of carrying heavier loads than the nonrigid, a semirigid airship could still be deflated and carried about the field.

As for the rigid-airship solution, Count Zeppelin claimed that the machine's internal structure of aluminum guaranteed that it could carry the heaviest loads at the greatest speeds over the longest distances without buckling. All three airship solutions were competing for a market that did not yet exist, except through the army. The German military, cautious about the count's wild technological claims yet aware of the progress the French were making in the field, decided to offer some funds to help him test his concept.[92]

Zeppelin built LZ 2 with newly hired help and incorporated improvements over LZ 1, including more powerful engines and an improved mobile-weight system.[93] Contrary to claims made to this day, the new machine had no vertical empennage or rear elevator surfaces to control altitude.[94] Small directional controls were placed on the lower fore and aft fuselage, but engineers did not grasp the importance of elevator surfaces until later. Instead, they kept the LZ 1 mobile-weight system, mounting the weight on a rail system installed between the two compartments and using a pulley to shift it. Furthermore, in view of the unhappy tendency of LZ 1's aluminum trusses to bend, technicians devised a new system of triangular cross sections for each support.[95]

Count Zeppelin's new overseeing engineer, Ludwig Dürr, initiated many of these changes. Fresh out of technical school, the twenty-one-year-old Dürr had joined the count in 1899 and remained in his service despite the latter's financial troubles in the wake of the LZ 1 failure. A calm, resolute personality, Dürr was the count's technical counterpart in the airship obsession. He would eventually rise to the position of chief engineer, yet he remained in the shadow of his more public (and vocal) employer and his successors.[96] Many early Zeppelin innovations were the result of Dürr's endless tinkering with metal formulas and structural shapes.

Despite Dürr's improvements, LZ 2 did not last. Following its first liftoff on 17 January 1906, the airship's engines thrust it up to 40 kph, fast enough to counter a moderate wind, then they quit. A successful landing ensued, but gale winds dragged the machine along the ground and destroyed it. LZ 2's demise did not cause any outcry. The only thing spectacular was the size of the wreck, which confirmed critics' claims about the flaws of the rigid-airship solution. Experts agreed with the count that a machine with a long flight capacity was desirable for reconnaissance purposes, but since big machines also required big bases and could not be deflated, it might be preferable to have a flotilla of smaller machines that could operate closer to the battlefield. Some engineers familiar with Zeppelin's project felt that it did, however, offer a "contribution to science" by confirming such theoretical calculations as air-drag coefficients and weight-lift ratios.[97]

The destruction of LZ 2 had two implications, one for engineering, one for Count Zeppelin's hopes of military success. The failure of the engines brought back to the fore some of the paramount problems of early aviation. The troubles of early engines included the weight of the machinery and the amount of fuel required to operate it. Altitude also affected fuel injection, reducing motor efficiency. The use of aluminum parts to lighten engines was still an uncommon practice, and technicians often cautioned against using aluminum for this purpose because of its unpredictable performance. As one engineer joked dryly, it was not just a matter of reducing gasoline consumption; when an engine quit in the air rather than on the ground, "attempting repairs was out of the question."[98] While these issues affected all airships, LZ 2's engine failure had special implications for the rigid-airship system.

The end of LZ-2 also dashed its creator's great hopes for a military contract. The German army turned away from the count's formula, authorizing instead the manufacture of inflatable nonrigid and semirigid airships.[99] This proved to be only a temporary solution, however. August von Parseval's nonrigid design, tested in 1906, was a disappointment to the military (it offered a very limited payload and range). However, it fascinated the public because of the number of flights it carried out without incident. By the summer of 1906, this and news from abroad (especially France) had once again turned aeronautics into a matter of popular interest. The kaiser's creation of a Society for the Study of Powered Flight—a response in part to a perceived "airship gap" with France—also seemed to confirm that the air age sweeping Europe had now come to Germany.[100]

Nationalist revivals were also sweeping Europe. France's balloon exercises always took place near the border of German-occupied Alsace and Lorraine, now a focus of resentment in the wake of the lost Franco-Prussian War. The

Dreyfus Affair, in which a French captain of Jewish descent was singled out and wrongly accused of spying for Germany, had run its course, leaving supporters and opponents of the officer's pardon agreeing on but one thing: Germany was the enemy. Worrisome to the Germans too was the visit of King Edward of England to Paris in 1904. Never had an Englishman been so well received in France. The kaiser's cousin now presided over the Anglo-French entente cordiale. This resounding slap in the face came in response to Germany's ongoing naval armament program, which threatened Britain's dominance of the seas. Old uniforms were dusted off, and while sabers were not yet being rattled, they were being polished—and the count was intent on forging an entirely new one.

Convinced anew that there was unfair bias in the government's evaluation of his work, Zeppelin demanded that the Imperial Chancellor's Office have his machines reevaluated by an observer from the Interior Ministry. This kind of bureaucratic jostling might have helped short-circuit the army's opposition to Zeppelin's work, but the observer turned out to be Major Gross, who had obtained an army contract for his own semirigid-dirigible project.[101] Gross's evaluation was, predictably, lukewarm. But for that very reason, it fed the myth that Zeppelin was attempting to foster—the myth of himself as a victim of government opposition—and thus ultimately helped build public support for his work. The count pressed on and raised additional funds. More successful this time, he was able to build a third airship, which, when it successfully flew in October 1906, attracted considerable attention.

In design and accomplishments, LZ 3 seemed to offer the "proof of concept" regarding rigid airships. Scientists quoted in the press claimed that LZ 3's flights represented "a glorious step, which the world will remember."[102] The technical improvements of the design were indeed remarkable. In hindsight, LZ 2 had served as the proverbial trial balloon, providing substantial information for LZ 3's design.[103] Improved versions of Daimler engines with better fuel distribution propelled it to a record speed of 54 kph. The new airship offered better directional controls and sported an added empennage based on designs that had first appeared on French airships.[104] This improvement made altitude changes easier and rendered the rolling carriage between the nacelles superfluous. Following early trials, Zeppelin engineers made their own further changes, adding several more directional and altitude winglets to the tail and so reducing the need for extra ballast.

The results were so promising that several commentators confidently stated that the airship would soon become a regular means of transport.[105] The count's only child, his daughter Hella, once flew on board, as did the German crown prince, suggesting that new levels of safety had been attained. Instead of the usual crew of four or five, one flight carried as many as ten souls. Endurance approached eight hours, and the ship even left the vicinity of Lake Constance to test its rudders under different atmospheric conditions (such as cold forest air).[106]

Upon witnessing the first flight of LZ 3, Major Gross reversed himself and recommended a grant of half a million marks to support the count's work. A second lottery, authorized a month later, brought in limited funds, but the following year the German Parliament allocated more money to allow the construction of a fourth airship.[107] If natural winds had once hampered the count's experiments, military, political, and international ones now favored them.

Such success drew public attention. Popular interest, although limited until 1907, began to increase thanks to media reports about Count Zeppelin, to the point that one politician felt it necessary to temper the awakening enthusiasm: "The judgment of the layman alone cannot decide matters; rather, expert opinion must decide." Echoing this point, the Interior Ministry declined to provide additional funding until the next machine flew.[108] The press, however, interpreted the government's show of caution as an excuse—a way to avoid making a financial commitment to the count despite the obvious national benefits of supporting his work.[109] Proposals for popular funding began to appear, independent of the lotteries the count had held to raise money. The Spandau City Council, for example, suggested that with a 10-pfennig donation from each German, over 6 million marks could be raised to keep Count Zeppelin's efforts going.[110] Similarly, Tübingen student associations took it upon themselves to honor his accomplishments, albeit with limited donations.[111]

The year 1907, then, proved a threshold to a new air age as reflected in both technical improvements and the general awareness of flight technology. Aeronautical associations reported notable increases in membership, while popular-science commentators discussed the potential dangers of air travel in terms comparable to those used for the automobile, which was wreaking havoc at the time.[112] On a lighter note, a Berlin paper philosophized that since balloonists had won a recent long-distance race against cars, practical flying for all was but a matter of time. Personal airships would replace the automobile, delivering everyone swiftly to his destination; the only problem might be finding one's way back from the pub, but "risks always accompanied innovations."[113]

Humor was but one indication that Germans were coming to terms with the new air age; novels were another. After the initial wave of futuristic novels in the late nineteenth century, a new kind, reflecting increased international tensions, appeared. One of the first and most widely imitated of these was a futuristic story of the Berlin-Baghdad link written by Rudolf Martin, a government official.[114] Published in 1907, the novel started with the much discussed railway corridor under development at the time but went on to imagine land-based travel giving way to an airship fleet, allowing Germany to extend its influence and claim additional colonies. This utopian vision coincided well with the expansionist designs of the German Colonial Society and the Pan-German League. Martin himself acknowledged and later emphasized the imperial potential of the new machine,[115] but in this novel he stressed its contribution to business, politics, and, especially, culture. Martin postulated that Germany's future flight technology would surpass that of England's sea fleet, although he was magnanimous: Germany would also help England regain territory lost to Russia and extend British holdings further into the sky.[116]

Martin asserted the primacy of Germany as a matter of course, but he stressed the wider significance of the new technology as a means of transport. Acknowledging popular resistance to change, Martin remained optimistic. The airship would overcome obstacles, much as the railway had done some eighty years earlier. If the "engine in the air," as he called it, lived up to its promise, it would have the effect almost of a new natural law, freeing humankind from gravity and shrinking distances more than the railroad had ever done. Unlike many German ultranationalists, Martin saw no conflict between culture and civilization, and he used the terms interchangeably:[117] the smaller the Earth, the greater the cultural strength of the new era in which distance would no longer hinder civilization. The opinion echoed the geographer Friedrich Ratzel's theory that the size of the state grew in step with its culture.[118]

Press coverage of Count Zeppelin's activities at this time often took the opportunity to reflect on the significance of such newly emerging technologies as electricity, the telephone, and the automobile. Popular fiction addressed such themes as well. Martin's novel, for example, pointed out that modern culture was distinguished from antiquity by the extent to which concepts of movement had been rethought.[119] In the case of the flying machine, the restrictions were of an entirely different order, and thus required new approaches to issues that the ocean liner or even the railway had not raised, ranging from questions of airspace to the assimilation of the new machines into the popular imagination.

Many commentators shared Martin's view that the airship was destined to break old barriers and probe new frontiers, and they expressed these sentiments even before Count Zeppelin's invention became a national symbol. In a very real sense, Martin and other aeronautical enthusiasts prepared the ground for the appearance of a real airship; tapping into the public fancy by means of fiction, they helped the airship reach unprecedented levels of popularity.

Collapse and Recovery

In 1907 the slow pace of German aeronautical advances contrasted sharply with progress in France, where the army's *Patrie* dirigible was setting endurance records.[120] Concerned observers now regarded Count Zeppelin as Germany's best hope for catching up with France, since the German Society for the Study of Powered Flight, established the previous year, had made little progress toward deciding which technical endeavor to support. A few designs, such as Parseval's, had undergone testing in a special wind tunnel built with help from the AEG industrial concern, but practical applications of the scientific knowledge thus gained remained limited.[121]

Much, in fact, depended on the armed forces, the only branch of government that might effect any quantitative or qualitative change in the progress of aeronautics. At the time, General Karl von Einem, who oversaw aeronautical matters for the military, had appointed a commission to study the different projects in existence. Although lukewarm, the commission nonetheless recommended an initial payment of 500,000 marks to help the count start construction, but it was routed through the Interior Ministry rather than the War Ministry.[122] The press interpreted this as a first: the imperial government was at last actually supporting an applied motorized flight project for "commercial and generally cultural purposes."[123] Although commentators acknowledged the role of military support in a nation's technical progress, they tended to emphasize the "inventive spirit" that had put France ahead in aviation matters. This did not stop one Reichstag member from stressing military courage and publicly wishing Count Zeppelin the same success he had earned fighting in the Franco-Prussian War.[124]

Significantly, the Reichstag members who had supported the count's funding came from both ends of the political spectrum, a sign in itself of the fascination the old man was beginning to exert.[125] By the time Count Zeppelin initiated his famous summer 1908 flights, the public was clearly well aware of his

LZ 4 leaves its floating hangar. In this June 1908 picture, the airship's tail does not yet have widened directional flaps. Onlookers around Lake Constance were fascinated by the floating hangar and what it contained, and often rented boats to go observe activities there. Here, a Swiss shuttle boat passes the hangar on its way back to Rorschach. The human figures atop the floating hangar give a sense of the dirigible's enormous proportions. Author's collection

patient efforts and of the controversy surrounding his funding.[126] This fact also helps explain the surge of public enthusiasm that occurred, taking the German government by surprise, when the count test-flew his fourth machine.

As part of a major effort to obtain further government support for his endeavors, Count Zeppelin agreed to carry out a twenty-four-hour endurance flight, which he hoped would convince everyone of the value of his machine.[127] However, LZ 3 was too small to accomplish such a flight, and it was placed in storage.[128] Instead, using the limited government funds that had already been awarded him, Zeppelin initiated a new project in November 1907.

The new LZ 4 offered further fine-tuning opportunities to the Zeppelin engineers. Between the two nacelles, where the weight carriage had appeared on LZ 2 and a gangway on LZ 3, a small observation and fuel-storage cabin was installed. Technicians also added two small rudders, one on each tip of the airship, hoping to make turns more effective. More powerful engines were installed, to increase cruising speed and to extend endurance up to twenty-four hours, the length of the required test flight.

On its twenty-minute maiden flight on 20 June 1908, LZ 4 rumbled about in uncontrolled circles, indicating that the tiny tip rudders were useless. The forward tip rudder was removed, and the rear one was replaced with a larger version, supplemented by parallel rudders on the ends of the horizontal stabilizers. The second flight, three days later, showed some improvement, but engineer Dürr ordered an increase in the size of the main rear rudder. The third flight, on 29 June, yielded good results. All the while, the press had kept the public informed of the technical developments; it was rumored that the twenty-four-hour flight was about to occur. However, Count Zeppelin first ordered a test flight half this length.[129]

On 1 July 1908, LZ 4 conducted its first long-distance flight, taking a route over Switzerland. The event attracted considerable attention; the crash of a German semirigid airship, the M 1, on the same day went almost unnoticed.[130] The flight broke all records, not only for altitude and endurance (it lasted over twelve hours) but also for the amount of public attention it attracted. Witnesses on the ground did not just see a 136-meter-long machine moving against the wind, they could count ten people on board, suggesting that this was indeed a proper means of transportation. History's previous aeronautical successes, achieved in flimsy-looking airplanes carrying one or two fliers, could not top such an image. The Zeppelin machine, until now just another flying contraption, passed into the realm of the sublime, its sheer size and distinctive sound inspiring a mix of fascination and fear. Two days after the Swiss flight, the king and queen of Württemberg took a brief ride over Lake Constance, thus implicitly declaring the machine "socially acceptable" and acknowledging that Count Zeppelin was more than an old eccentric.[131]

As more test flights took place, public fascination increased exponentially. The events of 1907 had planted the seeds of awareness in the European popular psyche; the summer of 1908 saw them flourish. In Le Mans, France, the Wright brothers showed off their perfected machine in a demonstration that left all other airplanes far behind. News of their success also stunned British observers, desperate to catch up with the Americans.[132] As for the French government, it announced the beginning of aeronautical maneuvers with one of its military dirigibles.[133] The "aerial mood" naturally spilled over into Germany, where press updates kept everybody informed of Count Zeppelin's preparations for his twenty-four-hour challenge flight.

LZ 4 took off from Friedrichshafen for Mainz on 4 August 1908. The distance, flying via Strasbourg before returning to base, was roughly 1,000 kilometers. Liftoff with Count Zeppelin and a crew of ten proceeded smoothly at 6 A.M. As the flight proceeded, a media frenzy developed, flooding press rooms with communiqués and leading to the posting of regular updates on city walls. The phenomenon culminated in what one historian has aptly termed an "imagined moment of idealism," when expectations surrounding a public event caused a suspension of individual differences.[134]

But the flight did not go as planned. Engine trouble forced a landing to effect repairs in the village of Oppenheim, a few kilometers from the return point. Far from putting a damper on public interest, however, the incident attracted even more attention to the flight. The writer Carl Zuckmayer recalled the dashed expectations of those who had hoped to catch a glimpse of the inventor and his machine, conflated into a single entity, "he":

> All of Mainz piled up on the Rhine riverbanks . . . for he was following the river's course. But he did not come. It got late and hot, and people became grumpy, for they had not eaten lunch, and they began to call airships and especially the rigid system unkind names. Of course, the wildest rumors began to fly. . . . When it was finally announced that he had been forced to land between Worms and Mainz . . . we youngsters jumped on our bicycles and rushed along the riverbank until, near Oppenheim, a police blockade held us back.[135]

As Zuckmayer further pointed out, this conquest of the air was a development comparable to the introduction of the automobile and of the use of electricity in cities. The airship, however, offered an even rarer form of entertainment. That evening, when it took off unannounced after repairs had been completed, making a surprise turn over Mainz "without the planned flag-waves and gun-salutes," all who were still awake stopped what they were doing and cheered the machine, some singing the national anthem.[136]

To ensure that this time the flight would be completed without trouble, six crew members were left behind in Oppenheim. But the return leg also brought engine problems; this time the forward unit quit, its crankshaft melted. By early morning it was clear that LZ 4 would not reach Friedrichshafen. Now the airship's other engine quit, and the machine reverted to a balloon, though a noble one, impressing onlookers from afar—it was at over 1,000 meters—but with a crew deeply concerned about where to land.

The crew vented hydrogen, and LZ 4 landed near Echterdingen. A military compound supplied men to help tie the machine down and guard it, since the news of the impromptu landing had spread; within a few hours, some fifty thousand people had gathered to gawk.[137] Most of the crew had left the ship, and Count Zeppelin, driven to the Hotel Hirsch, which doubled as a post office, was sending telegrams to his offices in Stuttgart and Friedrichshafen. Rumors spread that the count was at the Hirsch, and a new crowd formed there, prompting him to address onlookers from a first-floor window and thank them for their support.[138] As he concluded, a spontaneous singing of the national anthem erupted, leaving all present teary-eyed. Count Zeppelin returned inside to rest; there, a few hours later, he learned that tragedy had struck.

That afternoon, weather conditions around Echterdingen had deteriorated rapidly, and a storm broke out. Undeterred, spectators flashed open their umbrellas and looked on incredulously as thirty men struggled to keep the dirigible anchored. A strong wind lifted the machine from the rear and began to drag it along the ground. Stunned spectators began chasing it, as if hoping to corner it. The combination of ground proximity and stormy conditions likely caused a buildup in static electricity, and suddenly sparks flew, and the ship hit a tree. One of two men still on board jumped out upon hearing screams of "Fire!" behind him, while the other sought to vent the remaining hydrogen. Suddenly the whole ship was aflame, and the remaining crew, thrown off by the explosion, ran.[139]

One of the mechanics and two onlookers, wounded by the explosion, were taken to Stuttgart, but no one died.[140] The ship, however, was gone. Nothing was left but a smoking pile of metal debris to commemorate Count Zeppelin's failure. The crew were keenly disappointed, although the count, inspecting the crash site, kept his composure.[141] However, subsequent events turned this funeral pyre into a seminal moment in the apotheosis of Count Zeppelin.[142]

The accident, far from being the end of Count Zeppelin's experiment, proved instead a catalyst to further development. In a sudden surge of public interest that many were at a loss to explain, a spontaneous popular fund-raising drive raised over 6 million marks. Within a few days that development had prompted a stunned imperial government to award further credits (which it had promised in case of success), despite the mission's failure. Adverse circumstances had actually strengthened the haphazardly laid foundations of airship technology and its associated culture. The consequences would affect both German aeronautical development and popular culture.

Fascination with the conquest of the air, which had started as a ballooning craze in the late Enlightenment, ebbed and flowed throughout the nineteenth century as both a technical-scientific endeavor and a source of public entertainment. Developments in ballooning inspired prophecies of dirigibility and fantasies of castles in the air, which encouraged inventors to try their luck and skill at solving the puzzle of directed flight. The path to technological success was neither straight nor easy, but even the failures accustomed the public to the notion that human flight, although not yet a reality, might still be achieved. Even though the popular response to this notion was sometimes negative, the increasing awareness of it may have helped shift German opinion in Count Zeppelin's favor when he came onto the scene.

Zeppelin's endeavor unfolded in a setting of ambivalent attitudes toward science and technology. The count desperately needed the support of technical and scientific groups to convince both the public and the authorities that flight was not just a matter for eccentric inventors. But even successful flight offered no guarantee of support, as the count's first experiment showed. The right set of circumstances, in this case a mix of public fancy and official interest, had to coincide, and no amount of control or coaxing from the inventor could change this fact. This combination of circumstances finally came about in 1908. It was in that year that the development of European aviation, after a lull at the end of the previous century, reached its apogee. The novelty of flight at the dawn of a new century and the successes of other nations in the field, combined with the personal path of Count Zeppelin, were essential ingredients in the formation of the Zeppelin mystique, and of a dynamic new symbol: the Zeppelin airship.

The growing sympathy for the count and his airships also reflected a shift in the public perception of technology. The perceived curse of industrialization had led some commentators to question the very nature of progress.[143] Yet the "rehabilitation" of technology—the public acceptance of it—occurred much more quickly in the case of the count and his machines than it had with other recent developments ranging from the telephone to electricity. The inventor and his airships were seen as a blessing of the modern age and a manifestation of the technological sublime. It did not matter that flying machines owed their existence to the curse of industrialization. Now was the time to set aside such concerns and welcome this new phase of progress along with its herald. Count Zeppelin's moment of fame seemed to have arrived, and his invention belonged to the German public.

CHAPTER 2

The Machine above the Garden
Airship Culture in Imperial Germany

OFFICIALS THROUGHOUT GERMANY DID NOT expect the sympathetic public reaction to the crash of LZ 4 in August 1908. Count Zeppelin had failed to complete the endurance challenge, but as far as the public was concerned, it did not matter. "Anyone who, three months ago, heard nearly every day that Count Zeppelin was the biggest fool in Germany . . . is now told that he is the greatest German of the century," remarked an imperial administrator.[1]

What had seemed to be at most an interlude in a hot summer now spun out of control. The "popular spirit"—an idea inspired by Herder's *Volksgeist*—that social commentators had sought to define for the young united Germany was taking the unlikely shape of an old man and his flying machines. Ordinary Germans sent him enough money to build several new airships, while Reichstag parliamentarians argued about who had given Zeppelin his full trust first. Although the excitement never boiled over into actual social unrest, there was a hint of revolutionary fervor in the broad, cross-class basis of the count's popularity and the patriotic rhetoric associated with it. This broad base also offered an opportunity for public rituals that would merge old traditions and define new ones.[2] Naturally, as with any nationalistic trend, exclusions were bound to occur. Ironically, in this case the rituals tended to challenge Prussian leadership: a marginal, retired member of the South German elite had offered an apparent alternative to Prussian nationalism.

"Zeppelinism" and Its Expressions

The rise of engine technology and air power changed the European social scene between 1908 and 1914.[3] While the impact of Count Zeppelin's inventions did not compare with that of the automobile, it was nonetheless one of the factors that modified the cultural order. The "Zeppelin sublime" acted as a salute to a new epoch,[4] a manifestation of the cultural dimensions of Western aviation. Many nations experienced a similar moment before 1914, when a particular aviator, either successful or martyred in his exploits, inspired his compatriots to welcome and even seek out the new air age in a mix of patriotism and entertainment: Louis Blériot crossed the Channel for France, Lev Matsievitch died trying to break the altitude record for Russia,[5] and Count Zeppelin offered Germany new airships.

Observers of the Zeppelin phenomenon logically identified the public's support as essentially nationalistic, often ignoring its other dimensions. While contemporary commentators tended to overlook the military element of Zeppelin's appeal, that element gained in importance owing to the designs of the count and an increase in international tensions. This did not stop various kinds of nationalist feelings from expressing themselves through the emblematic power of the airship.[6]

As a unifying symbol that could represent the German nation without excluding other identities,[7] the Zeppelin was malleable to a variety of designs.[8] Stories about Count Zeppelin and his endeavors, retold and amplified many times, became means by which social and political groups could stress their participation in welcoming the air age and in supporting national ideals. Many different groups used the press to express their views on the matter, thereby influencing the layering of the airship symbol in mass culture.[9]

Some political parties, recognizing the count as an element of the new, multi-faceted German culture, tailored their official line to suit their followers' reactions to him. Social Democrats, for example, who before the crash of LZ 4 had cautiously supported the count's endeavors, distanced themselves from him immediately afterward.[10] *Vorwärts*, the Social Democratic newspaper, recommended that its readers not concern themselves with such "bourgeois" national issues as the airship crash. Only when the craze had clearly taken over the nation did the paper reverse its position and laud the count's work as a triumph of rationalism.[11]

Other organs took a more romantic view of the count's fate. He appeared to have made a Faustian deal and lost, in an adventure that followed the trajectory of romantic drama. Rumor and reportage kept the public informed, in novelistic installments, of the count's progress on a quest that was both utopian in its expectations and tragic in its outcome.[12] The parallel ends here, however. Rejecting the classical view that tragic endings satisfy the moral sense, Zeppelin's public felt that in this case hard work, duty, and obedience had not received their proper reward, as usually happened in the popular Grimm brothers' fairy tales.[13] A triumph over nature had occurred, and no one should deny it. The absurdities and contradictions of Count Zeppelin's situation now required reconciliation.[14] Although the crash appeared to be the end of the play, the final act, reports of the crowd encouraging the old man as he climbed into a car for the journey home, seemed to promise a new beginning for the Promethean hero Count Zeppelin had become.[15]

Interpretations of the incident varied, but the response was always the same: from schoolchildren sending stamps to shop owners shipping food, everyone felt a duty to help the count.[16] He had come to be seen not only as a rebel of sorts, refusing to accept defeat, but as the personification of visionary individualism. People everywhere felt compelled to recount their own highly personal reactions to the crash. One man recalled that his wife cried as if she were in mourning, while another thought of postponing his daughter's wedding festivities until after the fund-raiser.[17] Such responses were often interpreted as a breath of fresh air amid the "crass materialism" of modern culture.[18] A deep well of feeling had been tapped, to be expressed in multiple ways around the person of the count. His tribulations inspired not only a flurry of spontaneous giving but also thousands of poems, songs, and marches. Most were of dubious quality and quickly forgotten, yet they all attempted to use new words, images, and metaphors to define and explain the feelings Zeppelin had inspired.[19]

The myth launched one summer day in 1908 had grown to airship proportions. Notices from the Society of German Engineers denying its purported opposition to the count in the "heroic years" went unnoticed, overshadowed by

Zeppelin brings good fortune, literally. This interpretation of the spontaneous fund-raising drive of 1908 that saved Count Zeppelin's enterprise suggests that the nobleman should share his good fortune. The rhyme urges, in substance, "Zeppelin, the fund-raising sent you flying again / Now let some of your generous ballast rain!" ASTRA the lone-hero image.[20] At the center stood the count, embodying the quintessential German spirit. Despite his military background, many identified him as a man of peace, whom God should protect and whose "silver fish" swimming in the air were a sign of divine providence. The enthusiasm represented the gamut of patriotic beliefs, spanning the entire political and social spectrum. Few writers or politicians could make such a claim to universal cultural approval within a unified Germany.

The flood of support swept aside the few skeptics who challenged the adulation. Shortly after the Echterdingen crash, Emil Rathenau, the chairman of the AEG concern, proposed that an oversight committee representing government and business interests be established to manage the money obtained through the voluntary fund-raising.[21] Many such groups, composed mostly of middle-class elements, had gathered in cities to ensure an orderly channeling of such funds.[22] Rathenau's idea was simply to extend this oversight to the imperial level. However, the proposal became an issue of outside interference into regional affairs, raising questions about freedom and surprising actors on all sides. In the process, dormant anti-Prussian sentiments resurfaced. "We do not want control," clamored editorials in response to Rathenau's proposal, reminding everyone that the proceeds of the popular fund-raising drive conducted in April 1885 in behalf of Chancellor von Bismarck had reached him without any governmental interference. Reflecting the popular trust in Wilhelm II himself, another columnist claimed that the kaiser would not condone Rathenau's proposal, and even a German writer to the Swiss *National Zeitung* signing "on behalf of many" dismissed as "groundless" the arguments in favor of a central committee.[23]

The public dismay at the proposal expressed several strands of discomfort. To some, Rathenau personified capitalism seeking to steal the fruits of hard labor. To others he was the embodiment of soulless bureaucracy poised to corrupt the genius and the dreams of Count Zeppelin. Those on the left feared that if the government controlled the development of the airship, it would become a tool of Prussian militarism—a point of view that conveniently ignored the fact that Zeppelin himself advocated its military use.[24] The journalist Theodor Heuss—who in 1949 would become the first president of the Federal Republic—noted that it made little political sense either to support or to oppose the airship on the basis of its potential as an instrument of military aggression. For Heuss and many other commentators, all that mattered about the airship was its potential to achieve the dream of flight; its possible military applications, for good or ill, were far less important.[25]

The displays of patriotism that occurred, such as those seen in Berlin during the airship's visit in August 1909, suggested a high degree of civic unity.[26] Some observers nonetheless expressed discomfort about what appeared to be less a display of mass patriotism in response to a sublime machine than the development of a cult of personality around Count Zeppelin. Emil Rathenau's son Walter, an early supporter of the count, now had trouble accepting such blind adoration. He wrote to Prince von Bülow:

> At the time when both the mass of the people and those who form our popular opinion were equally decisive in their rejection of "the greatest man of the century," I, who had nothing but admiration for the count's patient effort and self-sacrifice, was already in the happy position, with my colleagues of the Society [of German Engineers], of being able to protect his life work. But as an engineer I find myself unable to join in the riotous acclamation of the assembled populace in Berlin—and even more in the kind of hysteria that now tries to put into the shadows the achievement of such great poets and thinkers as Germany could produce in Bismarck's day.[27]

Rathenau's comments expressed the liberal dismay that an aging soldier, neither engineer nor writer, could come to epitomize all things German. Perhaps there was a certain shallowness to the Zeppelin symbol, but this lament nevertheless overlooked the fact that the count and his machines belonged, in effect, to all Germans from Hamburg to Munich, and that Berlin could do nothing about it. A caricature in the satirical magazine *Simplicissimus* suggested the irony: instead of paying homage to the emperor, a humanized airship in evening dress presents its buttocks to Prussian power.

After the first wave of adulation, the count's popularity increased beyond all imagining. German reference publishers scrambled to keep up with the change. The German *Who's Who*, which did not even mention the count's experiments in 1906, tripled the size of its entry for the 1909 edition. Meanwhile, the popular Meyer *Lexikon* took advantage of a scheduled supplementary volume to catch up with the times and list the count's achievements.[28]

The count's ongoing success was the stuff of mythological heroes. Like them, he responded to a call to adventure and journeyed with the help of a few allies (his engineers, mostly). He battled strong opposition, descending into the depths before being reborn and achieving apotheosis as a quintessential genius

Zeppelin II auf der Fahrt nach Berlin

Erwartung unter allerhöchst fachverständigstem Vorsitz.

Gesteigerte Spannung unter allerhöchst bitto.

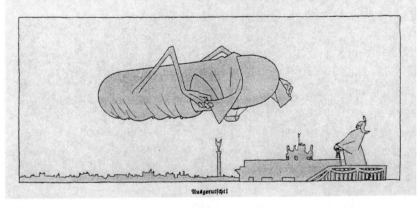

Ausgerutscht!

A humanized Zeppelin in coattails moons Prussian power. This 1909 caricature from *Simplicissimus* reflects southern and western German resentment of Prussia, a feeling based on centuries of feuds and exacerbated by Germany's unification in 1871 under Prussian leadership. The caricature coincided with the first airship visit to Berlin and pokes fun at imperial etiquette as the airship steps out of line. ASTRA who could do as he pleased.[29] As a "hero with a thousand faces" he reintegrated a grateful society.[30] The national reverence heaped upon him through celebrations and publications on his seventy-fifth birthday attests to the devotion that he commanded. The devotion bordered on hysteria. Siegfried may have killed the dragon in the forest, but Zeppelin had tamed Aquilon, or so the mythologizers presented it. More rational comments affirmed that the count rivaled Bismarck in worldwide renown, thanks to his background. Both of Germany's religious groups thus claimed kinship with him.[31] Although a Protestant, the count had grown up in a deeply Catholic southern German area,[32] and thus came to be seen as a crusader who could challenge the established northern Protestant Prussian order. Meanwhile, Protestant observers waxed poetic about the count as a representative of Martin Luther's work ethic and a preserver of his spirit.[33] While Zeppelin did criticize the system, however, he had not started a literal revolution. That he posed no actual threat to the imperial establishment made him acceptable to many social groups.

As a quasi-mythic figure, Zeppelin embodied a curious mix of simple and heroic virtues. He thus unleashed a frenzy of emulation.[34] Although not a "high priest," he was asked for his opinion on all manner of topics, and people sought to catch a glimpse of him the way fans today pursue pop stars. His gentle grandfatherly image, with the legendary white moustache and twinkling blue eyes (duplicated ad nauseam in caricatures and sculptures),[35] contrasted markedly with the obsessive determination with which he had fought the imperial bureaucracy. He seemed to accept as his due the adulation now heaped upon him, although in public he remained humble about it.[36] In his apparent ability to reconcile technology and nature, he epitomized a "new" German culture. Though a representative of the aristocracy, a class usually favoring preindustrial modes of thought and action, he had devised a machine essentially dependent on the products of industrialized society. His shift from a traditional role to that of a modernist of sorts probably contributed to the old man's mystique.[37]

While the count would not permit his name to be exploited for commercial purposes, he did lend his support to a few organizations and projects that in his view promoted positive images of Germans and Germany. Poetry and song collections intended for "school performances" and "patriotic celebrations" were acceptable. So were associations like the "Zeppelin Union," intended to promote "knowledge of nature, the use of nature's forces, travel, and excursions."[38] The "Jungdeutschland-Stuttgart," of which the count was honorary chairman, actively supported calisthenics as an activity to be performed in schools "in a patriotic spirit."[39] Gymnastics had played a considerable role in nationalist revivals throughout Europe, and it was thought that such activities might serve a similar purpose in a united Germany, especially under the count's patronage. To this end, various organizations claiming allegiance to Zeppelin's value system Germanized such pastimes as bicycling to give depth to a struggling national heritage.[40] His name became attached to any number of disparate causes. For example, some feared, in a reaction reminiscent of the Bismarckian *Kulturkampf* (the persecution of German Catholics in the 1870s), that Zeppelin's name might be used to promote a Christian world-view. Animal-rights activities, meanwhile, used his travails as a metaphor for their own, arguing that if the old man could succeed in the face of adversity, so too could the advocates of animal protection.[41] That the inventor and his airship became a means to so many ends is testimony to their emblematic power.

As German and foreign aviators set new records in airplanes, the count remained a towering figure. Orville Wright's visit to Berlin in 1909 and his later flight in an airship over Frankfurt became an occasion for celebrating both technologies, heavier- as well as lighter-than-air craft, rather than for making invidious comparisons between the two. "Zeppelin and Wright are the kings of the sky" waxed one observer, "[though] Zeppelin's name holds a special resonance everywhere."[42] Up to 1914 the count was the most popular personality in Germany, his fame extending beyond urban areas into regions "outside of the net of world transportation . . . which usually do not concern themselves with commerce or world events."[43] A visit from one of Zeppelin's machines built with public moneys was considered a mark of great honor, and invitations to fly to villages and small towns poured into Friedrichshafen every day.

There was nothing subtle about the charms of a Zeppelin visit. Whether intentional or accidental (schedules were often difficult to hold to), landings caused disruptions in everyday life and were occasions for impromptu celebrations.[44] Newspapers would print special editions when an invitation was accepted, and would print even more in case of a cancellation.[45] A canceled flight would cause an uproar in the press until a proper explanation was issued or a replacement flight scheduled. The attraction was such that independent companies offered charter flights in their own non-Zeppelin machines, hoping to capitalize on the craze.[46] But people wanted the real thing: a Zeppelin sublime.

As Count Zeppelin's associate Alfred Colsman described it, a "campfire atmosphere" gripped an area even before an announced airship showed up. Joy gave way to anger if lunchtime came without a glimpse of the airship.[47] The press had a field day, recording impressions that ranged from a child's perception of the count as a genie flying a liverwurst to an adult's desire to fly the airship to America to visit relatives who had emigrated there.[48] Even a simple fly-over might interrupt activities below. One report claimed that a funeral procession had stopped to sing the *Deutschland Lied* when an airship flew over.[49] Schoolchildren in Bonn were handed paper flags, as if in preparation for a state visit.[50] In Cologne, whose legendary carnival celebrations came to include many Zeppelin floats, rumors of a flight interdiction to protect military installations prompted a storm of protest in the press and forced the authorities to issue statements that would reassure the citizens: they too would see Zeppelins.[51] It was a matter not just of pride but also of keeping up with progress, and soon every city sought to have an airship station established.[52]

A Zeppelin visit symbolized a variety of parallel and sometimes contradictory hopes and developments, ranging from cultural pride to optimism in technology. In Leipzig, for example, the opening of the new airship station was staged to coincide with the hundredth anniversary of the battle that had sent Napoleon's troops into retreat. A commemorative flight was organized, and a few months later the German Airmen's League chose to honor the newly unveiled monument to the battle by holding its yearly meeting there.[53] Representatives from Austria-Hungary and Russia were also present at the celebration, suggesting that the flying machine and its inventor spoke to international interests as well as national ones.

The count's name quickly became interchangeable with that of his airships. However, the matter of terminology, of how to speak about the airships, was not entirely straightforward, as the following humorous dialogue suggests:[54]

> The wife: Now answer, husband, we could have easily slept a little more. Zeppelin is coming, but he does not keep his word.
>
> The husband, serene: Yes, little lady, you must recall that he is not coming but rather his *Viktoria-Luise*, a lady, is. And you know from your own experience: women can never hold to the appointed time![55]

The expansion of the railroad in nineteenth-century Germany had resulted in many new words being imported into the language (many taken from French),[56] and the same was happening now with the airship. How should one refer to this machine? In his patent application, Count Zeppelin had used the terms *Luftfahrzeug* (air vehicle) and *Luftfahrtzug* (air train) as distinct from

Luftschiff (airship), which denoted various flying devices.[57] Most publications came to refer to his machine as "the Zeppelin *Luftschiff*."[58]

As *airship* and associated terms made their way into the vocabulary, new words had to be invented, if only to differentiate lighter-than-air from heavier-than-air machines.[59] (This distinction was especially noticeable in Germany; German-speaking Switzerland and Austria continued to use the term *Luftschiffahrt,* "airship travel," well into the 1920s to refer to both types of flying machines.) The "technobabble" associated with early aeronautics blended older technical terms from the naval field with the poetry associated with open spaces. The vocabulary mirrored the dual effect of the nascent technology:[60] it offered mystery and charm yet relied on old wording to demystify the new dimensions it introduced.

German speakers were not the only ones grappling with what to call these flying machines. In Strasbourg, then part of the German empire, which LZ 4 overflew on its way to Mainz in August 1908, people alternatively referred to the machine as "le Zeppelin" and "la Zeppeline."[61] There was also some initial uncertainty about gender in German. While *Luftschiff* was neuter, the media remained confused about the gender of *Zeppelin.* Nonetheless, military and patriarchal associations soon made the term unequivocally male.

Meanwhile, language purists fought the use of imported aviation terms and sought to create a terminology not influenced by foreign languages. Since the unification of Germany, the ability to speak standard German without a regional accent had become a sign of education, and so had the use of technical jargon—but the jargon had to be Germanic. Heinrich von Stephan, who had inspired Count Zeppelin with his prediction of world airmail, set the tone by Germanizing all terms employed in the German postal system. More vocal in this endeavor, the General Association for the German Language used its influence to "clean up" words pertaining to flight.[62] The word for flier, for example, had to be *Flieger,* not *Aviatiker,* which was deemed too close to the French and English words. This orthodoxy informed the association's claim that "when you speak the German language, . . . you are a German."[63]

How German Is It?

Because of their proximity to Lake Constance, Switzerland and Austria-Hungary were constantly forced to deal with the symbol of German culture that Count Zeppelin's machine had become.[64] While both countries were linguistically linked to Germany, they had different ways of perceiving Germanic culture and its implications for their own identities.

Switzerland's first direct contact with Count Zeppelin's creation came in 1908 (although the Swiss press had covered his earlier work as well).[65] Balloon flight had by now become commonplace, but airplanes and airships remained an experimental technology coming from abroad. As part of a program to test LZ 4 prior to the twenty-four-hour endurance flight, the count and his associates initiated several flights that took the dirigible over the Swiss coast of Lake Constance, and eventually on a twelve-hour round-trip journey to central Switzerland.[66] The "Swiss Voyage," as it became known, stirred excitement in Helvetian territory. "In all the great Swiss cities we passed, there was jubilation and strong enthusiasm," telegraphed the airship's meteorologist Hugo Hergesell to Chancellor Theobald von Bethmann-Hollweg.[67] In Zurich, traffic came to a halt as people ran into the streets screaming and waving while the dirigible crew threw greeting cards overboard.

Emil Sandt, on board as a reporter, described in his articles the airship's passage over the "shining" Rhinefalls and Lucerne, the "pearl of the Swiss jewel box."[68] From above it was "a mix of romanticism and culture, of nature and civilization"—a spectacle "that no man before us has had a chance to enjoy."[69] Such descriptions were legion in accounts of airship travels, and they fit remarkably well with the romantic vision of Switzerland that Johann Ludwig Uhland, Johann Jakob Bodmer, and other German poets had created in the eighteenth and nineteenth centuries. Their dream of Switzerland as a model of cultural, religious, and political harmony may have strayed far from reality, but it had a lasting effect on the literature of the time, especially with regard to the depiction of nature.[70]

Sandt's description of the flight also had much in common with the work of another romantic writer, Jean Paul (J. P. Friedrich Richter). In his fictional journal of the balloonist Giannozzo,[71] Jean Paul argued that political borders were now pointless because they could be overtaken by air. The "small state" mentality typical of Germany would give way to a culture of open space and open minds. The idyllic Switzerland of the German imagination was an object of envy and an element of nineteenth-century German identity that led many to dream of retracing Goethe's steps there.[72] Many Germans now imagined flying over Helvetian land and seeing the young Rhine,[73] and the Zeppelin company would eventually capitalize on this fantasy by offering chartered flights to the Rhinefalls.

In Switzerland itself, the Zeppelin's visits inspired only a mild degree of pro-

German sentiment, and in fact they drew criticism from Swiss Social Democrats, whom their German counterparts quickly rebuked.[74] Some of the Swiss conservative press did, however, emphasize the German aspect of the achievement, implying a contrast with Switzerland's own difficulties with establishing a national identity in a multicultural setting. Generally, though, foreign papers noted that the enthusiasm expressed in Switzerland proceeded less from admiration for Germany than from wonder at the notion of flight itself. In Switzerland as in other places they later visited, Count Zeppelin's creations impressed would-be airmen with their sheer size and apparent stability.[75] The brief flight of LZ 4 over Lucerne also prompted the establishment of an airship station there in July 1910, out of which excursion flights took place two summers in a row on board a rented French airship.

Any fascination the Swiss may have had for Zeppelin's machine, then, had more to do with mastery of the air than with any desire to share in German cultural pride. In neighboring Austria-Hungary, however, the reaction was more ambivalent. Unlike their Swiss counterparts, Austro-Hungarian fliers were often able to draw official attention to their efforts. Although support for flying varied, Viennese media and society took an interest in the ongoing saga of pioneers such as Wilhelm Kress and Igo Etrich.[76] Airships were met with greater skepticism. The veteran balloonist and publisher Viktor Silberer often gave bombastic talks attacking the dirigible concept as unsafe. By the time of Count Zeppelin's spectacular recovery from the 1908 crash of LZ 4, Silberer had time and again made clear his view that the elements would always defeat the airship.[77] Other commentators, however, came to the count's defense. They included Hermann Hoernes, an Austro-Hungarian army officer and flight pioneer, who argued for the necessity of experimenting before deciding which means of transport was most appropriate.[78]

The Austrian press gave extensive coverage to the count's early efforts, in particular the 1907 flights.[79] As a result, by the time of the August 1908 crash his name was as famous in Vienna as it was in Berlin. The destruction of LZ 4 struck a chord of sorrow in Vienna, where the reaction was described as "a mourning in which the entire civilized world takes part."[80] Count Zeppelin's recovery from the disaster awakened Austrian interest in the possibility of using airships for transport and military purposes. The imperial army eventually funded several indigenous experiments and bought nonrigid airships from both France and Germany.[81] There was little chance of obtaining a "real" Zeppelin, let alone welcoming such a machine on a visit to Vienna.

Members of the Austrian government did not share in the popular disappointment over this fact. Here, behind the scenes, a different perception of the crash and its consequences was making headway. There was a growing sense that the German fund-raising drive had been an expression of aggressive nationalism. The Austro-Hungarian ambassador to Berlin attempted to convey a qualified version of this impression (whose source he preferred not to cite). The popular enthusiasm for Count Zeppelin, he said, may have been in some part symptomatic of warlike tendencies, but there was more to it than simple warmongering: "The popular feeling that honors him in every way rests, in my view, on much deeper perceptions than those that would be evoked by warlike urges or the hope for military expansion."[82] While the ambassador praised the count for his efforts, he worried about the exclusionary implications of the new airship age for the balance of power. Airship flights represented an achievement of the cultured German people, yet from the outside they also seemed to symbolize the power of Bismarck's united Germany, from which Austria had been excluded.

Most pro-German Austrians preferred to side with the airship enthusiast Rudolf Martin, who suggested that the airship could allow further new cultural alliances: "Let us assume that the German Empire and Austria-Hungary agree on joining in a federated state; such a state will then enjoy the advantage of closer influence on the Balkan peninsula and Asia Minor thanks to the use of its airships. These machines will prove useful at once, for even in the Orient, one will perceive the necessity of a close union with the expanded German Reich."[83]

Regardless of the mixed feelings the German airship may have evoked, Vienna, like other cities, hoped for a Zeppelin visit. Distance and altitude—the difficulty of passage over the Alps—precluded any early courtesy call, pushing back the date of the visit time and again. One periodical published a children's chant asking Count Zeppelin when he would reach Vienna.[84] The long wait may have accentuated a feeling of second-class citizenship vis-à-vis Germany, since other famed aviators did pay visits to the Austrian capital, including the Channel flier Louis Blériot, whose airfield demonstration Emperor Franz-Josef traveled to see.[85] The blow was softened, of course, by the fact that no country aside from Switzerland had received a Zeppelin visit, either.

All this led to conflicting attitudes about Zeppelins and their inventor. While acknowledging the German effort, publications also noted Austrian progress in airship technology.[86] Several Austrian pioneers had built airships but received

The airship as an expression of Germanness. In the years before World War I, Europeans had developed a community awareness that associated technology with nationalism. This Austrian postcard reflects such associations, linking the airship to German identity. ASTRA

little funding, since the army could not afford to operate any more dirigibles than the few it already had.[87] Austria's pride in its own builders remained strong, however, and their efforts were often compared directly with those of their German counterpart. Several reporters reminded readers that the airship pioneer David Schwarz, whom they described as Austrian (a Russian-born Jew, he was buried in Vienna), had hit upon the rigid-hull concept before the count had.[88] Another columnist called the airship pioneer Alfred Mannsbarth "the Austrian Zeppelin."[89] Germans were not altogether pleased by these comparisons. Several newspapers published miffed reports on how the Austrian press had belittled Zeppelin's contribution to the rigid-airship system.[90]

Eventually, Austria's long wait for a Zeppelin airship visit ended: the count flew the *Sachsen* to Vienna in June 1913, circled over Schönbrunn Palace, and landed at Aspern Field north of the city to joyful shouts of "Grüss Gott!" from the crowd.[91] The visit became an occasion for demonstrations of solidarity and sympathy, culminating in the count's visit to Schönbrunn Palace to receive a decoration from Franz-Josef.[92] Comments on the visit suggest feelings of admiration coupled with a certain wish to be a part of history in the making. Quoting the classics, the mayor of Vienna reminded everyone present that such an invention "means a new era, not solely for the people to which we belong, but also for the entire world."[93]

There were some complaints from minority groups. For example, the Jewish weekly *Die Wahrheit*, stressing the legacy of David Schwarz, complained that Jews who had rendered services to the state and to Germanness in Austria were being neglected.[94] Nevertheless, most of the commentary focused on the importance of flight and the mythic import of the new technology; the response throughout the multicultural empire was mostly one of admiration.

The excitement the machine generated as it flew over Vienna was a vindication, for good or ill, of Germanness. This fact appeared to be an essential component of the count's success. The city was determined to demonstrate to the pioneer that "German Vienna" could salute his wonderwork as enthusiastically as any other city he had visited so far. The circle over Schönbrunn was not just a gesture of mutual respect but a special political and national moment. Zeppelin's achievements were a triumph of human intellect, and by his side "the German people stood," wrote one Viennese columnist. He went on: "This is the result of action. One perceives the greatness. Can we be so bored as long as this joyous gratitude exists?"[95] Similar remarks brought up issues more closely related to the airship's potential as a war machine. This was the "flying torpedo" of which futuristic war novels spoke. The *Sachsen,* sitting on Aspern Field, was but a slow transport ship, yet the association seemed unavoidable:

> Count Zeppelin would not be a true German if, following his first tri-umphant parade to Berlin, the former cavalry leader had not underlined the [military] value of his airship. We rejoice sincerely, as close allies, that our friends may add such reinforcement to their strength, which gives them a great advantage in case of an [international] incident. . . . Our first and high-est sympathy goes to the great victory for civilization and culture that Count Zeppelin as an aviator has achieved.[96]

This commentator saw no incompatibility between war on the one hand and civilization and culture on the other. In the grandeur of mastering the air, such barriers were broken down. The Social Democratic *Arbeiter Zeitung* was able to separate the airship's war potential from its peaceful uses and came to echo its German counterpart *Vorwärts* in acknowledging the Goethean genius that had spawned such an invention. At the moment, capitalist chains might bind the miracle of technology, but the latter would soon turn into a hammer and break them.[97]

The visit of the *Sachsen* to Vienna demonstrated cultural but not political assimilation of the machine to Germanness. Numerous proud expressions of belonging to a common heritage appeared, but few called for the concrete po-litical uniting of the heritage. The count had described his visit as a means of strengthening already close ties between two German peoples.[98] The comment was ambiguous enough to fit either interpretation. The count had in his youth supported the *grossdeutsch* solution of a unified German state under Austria rather than Prussia, but following the Prussian victory at Sadowa in 1866, he chose Bismarck's side.[99] The nationalism issue in the case of Count Zeppelin's visit, then, was certainly cultural but did not extend to include such patriotic political parameters as an open association with German air efforts. Indeed, the week that followed the count's visit witnessed a large aviation meeting in Vienna intended to raise money for an Austro-Hungarian army air arm com-posed of airplanes, but not airships. In Germany, a similar attitude prevailed in the military, leaving Count Zeppelin in a bind.

Finding a Use for the Airship

Count Zeppelin had planned all along to sell airships to the German army. The public moneys obtained in 1908 could not sustain his enterprise for more than a few months. The public adulation masked the tensions that continued to reverberate in his relations with the imperial government. The airship crashes that occurred in Germany between 1908 and 1914 exacerbated these tensions and raised questions about the value of such a machine, to the delight of its opponents.[100] These accidents, along with new competition, forced the Zeppelin enterprise to diversify its production.

Had the crash of LZ 4 in 1908 been the last airship accident, the few notable critics of the technology, such as the journalist Maximilian Harden (who coined the term *Zeppelinism* to describe the craze), would have been silenced.[101] Continuing technical difficulties, however, kept the new machine from winning full acceptance. In April 1910 the German army conducted airship maneuvers near Cologne. Although initially viewed as a further occasion for rejoicing at Germany's new mastery of the skies, the event proved to be a new source of sorrow: heavy winds ripped LZ 5 (also known as Z II in the army) from its anchors and sent it crashing into a hill. Fearing that public opinion would no longer support him, Count Zeppelin and his associates blamed the army for poor handling procedures, hoping to deflect criticism.[102] The army brass grew upset at what they viewed as unjustified press attacks—an inquiry determined that the media had indeed misrepresented the situation—but Count Zeppelin was not out of the woods yet. Two more airships, LZ 6 and the LZ 7 *Deutschland,* used for civilian air excursions, were destroyed in the summer of 1910, with no loss of life. The military contracts dried up, and for two years Zeppelin desperately awaited new orders.

In the meantime, observers began to question the soundness of the rigid-airship system. Zeppelin technology made considerable progress between 1908 and 1914, especially in the matter of engines. The first motor installed on a Zeppelin, a Daimler, had had an output of about 15 horsepower and weighed over 385 kilos. A decade later the new Maybach motor—dissatisfaction with Daimler engines had prompted Zeppelin to switch to another supplier—had increased efficiency to 145 horsepower for a weight of 445 kilos.[103] In-flight turns improved considerably, but the new engine could not solve the problems of field-steering; because of its size, an airship on the ground remained an ideal target for winds. There were other types of airships, designs that withstood rough landings better, handled field maneuvers more easily, and required less sophisticated ground equipment than Zeppelin's machines, but they were equally vulnerable to wind.[104] Critics began to ask: Would not the millions invested in such white elephants be better spent on other aeronautical projects?

Until World War I the field of aeronautics encompassed a peculiar mix of legitimate aerial experimenters and manufacturers and a wide range of amateurs, from luckless idealists to complete crackpots. Some even traveled to Friedrichshafen and discussed the latest in airship matters with anyone who would listen.[105] Some of the ideas were not so far-fetched and might have improved on aspects of Zeppelin's designs. Others looked good on paper but ignored such elementary matters as weight problems associated with using steel instead of aluminum. This did not prevent some engineers from arguing that once applied science broke new barriers, it would also solve such problems.[106] A favorite idea, which persisted well into the 1920s, was the application of a so-called third way, combining lighter-than-air with fixed-wing technologies. The main advantage, its defenders claimed, was safety: if the balloon/airship component lost buoyancy, it could simply glide to the ground. With aerodynamics still in its infancy, many such ideas found their way into both the specialized and the popular media, intent on covering any concept that might bring new prestige to the airship's progenitor. However, the intricacies of a winged airship, even on paper, were such that none of the proposals made it past the drawing board.[107]

More rational projects led to the creation of new businesses, from competitors like the rigid-airship manufacturer Schütte-Lanz[108] to firms investing in related technologies. City councils fought for the construction of airship installations and testing grounds.[109] A contract for airship supplies, from hangar construction to food served on board, not only could keep a small company afloat but also could provide it with a prestigious form of advertising. Three factories shared contracts for airship envelopes, which were sewn out of either doped cotton or thin rubber sheets.[110] In the case of the German Continental corporation, use of the airship and ballooning symbols in its advertising became central to the marketing of its rubber products.

The infant aviation market was small and unstable. Zeppelin's competitors would have liked nothing better than to see his enterprise sink under the weight of its difficulties, but there was the risk that the whole nascent industry would be destroyed in the process: the count's invention had too much symbolic importance in the field. As for Zeppelin and his associates, they responded to competition by creating several new companies of their own. Between 1909 and 1912, the Zeppelin company underwent a social and cultural transformation at the managerial and technical levels. Although in the public eye everything remained centered around the persona of the count, the sheer size of

Continental **Pneumatik und Ballonstoff**

eroberte die Landstrasse und die Luft.

Conquering the skies and the roads. The Continental rubber company capitalized on its role as a manufacturer of products for airships to promote its automobile tires: quality products would ensure a successful trip, whatever the mode of transportation. This 1909 ad relies on a heavy dose of fantasy. Cars and airships alike were notorious for the haphazard working of their engines, and in both cases passengers had to don heavy furs or raincoats against the cold, dust, and leaking oil. ASTRA his endeavor required new management and design methods. The count kept a close eye on the engineering aspect of the enterprise until 1912, but two managers, Alfred Colsman and Alfred von Soden-Fraunhofen, had begun initiating the transformation of the company three years earlier.

Alfred Colsman, who would lead the Zeppelin concern until the late 1920s, offered an interesting contrast to Count Zeppelin. He had studied business and economics in Berlin and Geneva, and brought with him the outlook of a cigar-chomping captain of industry rather than of an old nobleman. First active in the family silk trade, he switched to metals when he married the daughter of the aluminum producer Carl Berg. He thus came to know the count through his father-in-law's involvement in the Zeppelin project during its early years. The respect he had for the count was mutual, which helped Colsman implement changes in the company structure starting in 1909.

Colsman's mission—and achievement—was to transform the Zeppelin company from a cottage industry into a full-scale concern. Physically, this meant moving the company grounds from its floating hangars to land purchased near the Riedle forest outside Friedrichshafen. When the military contracts dried up in 1909 after the delivery of two airships, the Zeppelin company found itself with a "white tail" airship: a machine built but not delivered. Colsman determined that the best alternative use for the machine would be civilian air transport between German cities. Count Zeppelin gave his grudging approval, and the DELAG airship transport line began operations that same year.[111] Colsman thus killed two birds with one stone: he kept the company afloat and ensured continued public interest by offering, for a hefty fee, the chance to experience firsthand the count's "gift to Germany."

Colsman also helped extend the Zeppelin company's involvement into other endeavors, including the Maybach engine corporation,[112] a Zeppelin insurance system, and a worker housing project, which took several years to complete.[113] In the latter case, the shift of the company to a fully paternalistic system reinforced the image of Count Zeppelin as a protective figure, but it also was based on sound business acumen: trained labor was hard to come by, and housing prices were such that in the early years of the company, many workers tended to leave the area rather than come to it because more affordable housing was available in large cities.

The worker-housing scheme did not mean that Colsman effectively controlled the company's workers or their unions. In March 1911, for example, following a drop in wages that came on the heels of several airship accidents, low morale on the factory floor prompted a Socialist-led strike. The strike was ended by a lack of unity among the workers' groups—there was a strong Christian metalworkers' association— rather than by Colsman's negotiating skills or Count Zeppelin's mournful calls for a return to work, but unrest continued throughout the year.[114] Yet Zeppelin managers got the message and sought to ensure greater loyalty by subsidizing additional housing units and supplying extra benefits. As both working conditions and the business climate improved, the number of job applicants increased, and the German public, barely aware of the labor unrest, continued to accept the company's corporate image of professionalism associated with high technology: to work for Zeppelin meant working for a symbol of quality.

At the engineering level, other Zeppelin associates helped strengthen the company's commitment to quality by effecting a shift to a new style of industrial practice.[115] To do so, however, they had to reconcile the old engineering culture with a new one.

Ludwig Dürr, the count's leading and longest-serving engineer, was a tinkerer. The early design of airships had been a process of trial and error, and that had given Dürr a feel for the machines. An extra flap on the tail, a new fuel tank in the fuselage—such additions had been tested in flight, adjusted by hand or deactivated on the spot in direct response to the machine's performance. But this artisanal approach to engineering could not work with a big design team: orally transmitted knowledge might be interpreted differently depending on the recipient's background and training. Engines, for example, continued to be a problem. Few could run for more than a few hours at a time, yet no protocol existed to deal with this situation. A freshly hired Claude Dornier, who would go on to design and build hydroplanes and bombers, recalled one such situation. Both engines of the airship he was on had just quit:

> The engine mechanic [in the rear gondola] kept trying to fix it when suddenly oil collected in the engine began to burn. We took off our leather coats to fight the fire. After a few tense minutes, we succeeded. In the meantime the forward engine had started again. . . . To my great dismay, my recently purchased leather coat had not fared well in the firefight. Only later did we discover that hydrogen leaking from the ballonets and remaining in suspension in the fuselage had posed an additional explosive danger.[116]

If Zeppelin personnel dealt with trouble this way, how could one train nonfactory crew to act systematically?

The Zeppelin engineers needed a body of knowledge based on scientific testing. Although several of Zeppelin's machines had indeed benefited from such experiments, the process of adopting improvements had been piecemeal. "Old Zeppeliners" maintained that if something worked, why improve the formula? But newly hired engineers came better prepared, and with a clearer sense of how to go about incorporating improvements. However deeply they admired the count, their admiration was balanced by a passion for technical work.

Such was the case with Alfred von Soden-Fraunhofen. What Colsman did for the Zeppelin company's business culture, Fraunhofen did for its technical side. His background was similar to Count Zeppelin's: born to a Bavarian noble family and home-schooled, as was the family tradition, he had gone on to serve in the military. But there the parallel ends. Fraunhofen went on to study electrical engineering and began working, thanks to help from the engine pioneer Karl Maybach, for the Daimler-Benz company. Bored out of his wits by the less-than-ideal working conditions, he moved to the MAN engine company in Augsburg, often journeying from there to Lake Constance, where he had enjoyed observing the flights of LZ 3.

In 1910, Count Zeppelin, who knew Fraunhofen from his own association with Daimler, approached him and requested his services. The count was thinking in particular about the need to develop an aeronautical test stand that could also be used to design airplanes, should competition in the field make diversification necessary.[117] Colsman oversaw the details, and within months Fraunhofen was heading the testing section. Zeppelin engineering culture now broke new ground as it moved toward systematic rather than piecemeal testing. Engineering statisticians such as Claude Dornier and Karl Arnstein had begun to turn up in Friedrichshafen, their calculations replacing the dozens of test flights older engineers like Ludwig Dürr had ordered.[118]

The media and the public did not take note of the shift in engineering and business cultures that was occurring in the Zeppelin concern; what counted was that there were airships in Germany. But as a result of this passion for airships, the German aviation industry lagged far behind France in airplane development. Although the War Ministry had steadily increased funding for both the airplane and airship industries between 1910 and 1913, by the end of that period the seventeen German airplane manufacturers had a combined capital of only 3.2 million marks, compared with the Zeppelin company's 3.5 million. During that period some twenty-four airships had been contracted for.[119] This was partly the result of strong lobbying on the part of the Zeppelin company itself, which waved the risk of bankruptcy in the face of government officials.[120]

Even though airplanes were making considerable progress in Germany, airships remained the quintessential German product. When speaking of airplanes, Germans continued to concentrate on foreign builders and fliers, giving short shrift to most German airplane pioneers. For example, press coverage of August Euler, bearer of German pilot license number 1, described as "dangerless jumps" his demonstrations of skill at the 1909 Frankfurt air show.[121] Until 1913 the media punctually noted the anniversary of the Wright brothers' first flight, and regularly mentioned the names of the Brazilian Alberto Santos-Dumont and even the Frenchmen Louis Blériot, Ferdinand Ferber, and Adolphe Pégoud. German achievement, when mentioned at all, was represented by the late Otto Lilienthal, a brilliant pioneer whose death in a gliding crash in 1896 had gone essentially unnoticed.[122]

Despite this shortsightedness in the matter of airplanes, German interest in aviation steadily increased thanks to the Zeppelin phenomenon. Even airship crashes were good publicity, reinforcing the notion of sacrifice for the sake of mastery of the air.[123] Public support for airships, initially a spontaneous—and thus unstable—factor, was eventually channeled into numerous aeronautical associations. Membership in such groups exploded in 1907, and three years later they had over sixty-five thousand members spread out among seventy-five groups.[124]

Other expressions of German interest in Zeppelins, and in aviation generally, included a heavy demand for consumer items sufficient to feed an army of small entrepreneurs. Photography had immortalized Count Zeppelin's exploits in countless books, and film soon followed suit. Those who lacked the means for a flight on the DELAG line, the company operating commercial Zeppelin airships, could afford to watch a five-minute movie in theaters, enjoying a bird's-eye view from the cabin of the *Schwaben* on its way to Düsseldorf, the champagne glasses on passengers' tables, and even the machine's shadow darkening the ground below.[125]

The airship provided such a rich visual experience in itself that, not surprisingly, the public's demand for souvenir items knew no bounds. The Zeppelin became a new means for marketing a variety of products, especially since German trademark law at the time did not forbid such activity.[126] This explains the multitude of items that appeared with the count's name or likeness on the label. While he authorized a few of these uses, like a sausage paté from a

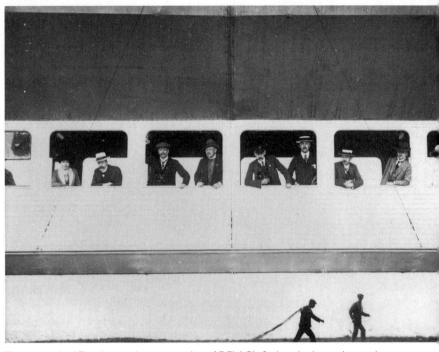

Train or airship? Travelers in the center cabin of DELAG's *Sachsen* look out the windows while crew below remove the ropes holding the airship down. Placed in service in 1913, the *Sachsen* carried out many charter flights over its namesake region during that year. Flying aboard an airship was a special experience that called for men and women alike to don their Sunday best. Author's collection

butcher friend (still sold today), he opposed most such practices but could do little to stop them. Already synonymous with German high technical culture, Zeppelin's name now also extended into popular culture at several levels. One manifestation of this phenomenon was the proliferation of Zeppelin-related kitsch objects widely derided by critics of the time.[127] The series of objects that accompanied the success of Count Zeppelin fueled new attacks, with such "leaders of good taste" as Gustav Pazaurek complaining about the undiscriminating affection of the masses for Zeppelin dolls and airship-shaped lamps. Zeppelin items evidently appealed to Germans of all ages: the Lehmann toy company, which started selling tin models of the Zeppelin airship in 1907, experienced an increase in sales within months of the Echterdingen accident, while competitors sought to appeal to male adults with miniature airship tape measures.[128]

The craze was also remarkable for its appeal to both genders. At first the airship community, for all the support it drummed up, was male-oriented. The only women working in Friedrichshafen were secretaries or seamstresses sewing up the cotton airship envelopes. Yet mass culture and light entertainment in the late nineteenth century was a traditional part of women's private sphere.[129] One of the early noncorporate attempts at channeling public support for Zeppelin had been organized by the wife of Zeppelin's lawyer Ernst Uhland, and soon other women were taking an interest in the subject. Airship clothing, an outgrowth of the balloon fashions of the Enlightenment, offered many new targets for ridicule, although a few fortunate women profited from the fashion's advocation of the wearing of pants.[130]

Well-to-do women naturally took part in Zeppelin flights strictly as passengers, as the count's daughter had done. Whereas a few German female airplane pilots broke the gender barrier in imperial times, female dirigible enthusiasts were less successful. Ballooning had become an acceptable form of flight for women, but only one woman succeeded in overcoming the opposition of the male establishment to qualify for a license to fly an airship, and even then it was only an honorary pilot's certificate.[131]

Anyone who failed in the attempt to become an airship pilot could always demonstrate courage by paying handsomely for a DELAG ticket. To be able to describe one's experiences above land conferred distinction in the upper spheres (literally) of German society. The happy few who experienced flight and recorded their impressions of it often described the joy of floating as a sensation they would gladly repeat if given the chance. The intricacies of flight technology were of no interest; what mattered was the new world-view one could enjoy while swimming in midair.[132]

DELAG executives understood this and sought to dispel any fears of flying by publishing tour guides reminiscent of Baedeker's. The fantasy of flight was all very well, but a certain appearance of routine was essential if people were actually to accept this new mode of transportation, especially in light of the spectacular crashes that did, at times, occur. Automobile accidents were far more numerous, and train crashes far deadlier, but even the dangers of railroad travel could not rival, for sheer terror, the fear of falling out of the sky.[133] To allay such concerns, DELAG operators noted in their brochures that all the deadly accidents that had occurred had involved military machines (they did not mention nonfatal commercial incidents). "If one always chose to be afraid, absolutely nothing could be undertaken," explained a press release at the opening of the 1914 flying season (no machines flew in wintertime). "A railroad car

The flying Pullman car. In an effort to demystify air travel and allay fears associated with it, DELAG circulated various brochures and illustrations intended to show how pleasant it was to ride in an airship. Champagne and cold meals were sold during sightseeing trips. This artist's view appeared in the British press and represented the cabin of the LZ 7 *Deutschland*. By the time the picture appeared in July 1910, the machine had been wrecked, albeit with no loss of life. ASTRA

can derail, an automobile can turn over, horses can go wild."[134] Of course, the best publicity remained the fact that a "typical German man" in his seventies regularly flew on board his own machines.

On the ground too the elite debated the benefits of airplanes and airships. Commenting on the potential benefits that aviation as a whole offered to humankind, the Nobel Prize winner Wilhelm Ostwald suggested that in the case of aeronautics the normally dehumanizing effect of technology might be reversed. The liberal family magazine *Gartenlaube* went a step further and remarked that the airship coincided with the overall perfecting of inventions that accompanied the advent of the new century. Such positive developments, "pushing forward the boundaries of progress," made it "a joy to be alive."[135] Borders were now meaningless, and "the new German Reich [which] began with the Customs Union" was just one example of what the new global mail service would accomplish.[136]

All flying machines, whether heavier- or lighter-than-air, had potential applications that everyone could appreciate, but the sheer mass of an airship, especially one of Count Zeppelin's, made it particularly appealing. What distinguished the count's creations from other flying machines, beyond their amazing flight capacity, was the aesthetic pleasure they provided. A huge tube with noisy engines, a Zeppelin airship nevertheless gave the impression of gliding effortlessly through the air, immune to gravity. A product of the Industrial Revolution, it nevertheless suggested artistry and craftsmanship, and contradicted the dreary image of mass production. In effect, it represented a kind of reconciliation to the modernization process by allowing a dream to come true. It was a device fit for the "First Machine Age," the age in which, as young Futurists like Filippo Tommaso Marinetti predicted, men would become free to control their own destiny through machines.[137]

Standardization and anonymity in technology was a subject of hot debate in the European art world of the day.[138] Artists fought to cast technology as a creative endeavor,[139] but they suspended judgment when considering Zeppelin's machines. (There were, of course, exceptions to this view, as when the artist J. L. Forain described one ugly new modernist building in Paris as "the Zeppelin on the Avenue de Montaigne.")[140] Generally, Zeppelin's airships were described as smooth, symmetrical, and stable—a unique combination of artful design and functionalism that even the airplane did not offer at the time.

The product of rational thought, airships nevertheless served a romantic notion of floating in the air. They were an exception to the rule, a kind of technology that did not subjugate the individual or force him into anything but rather granted him an opportunity to expand his horizons both imaginatively and literally. Although skeptics like Walter Rathenau felt that technical inventions could not be complete and perfect, they agreed that such devices could be beautiful and inspiring.[141] The airship's beauty allowed each observer to interpret its meaning according to his own political and aesthetic preferences. Nationalist pride blended romantic visions of the airship with romantic notions of territory: "The tools of our aviation are elements born in our cultural milieu

Zeppelin advertising: the future of milk processing. As a herald of modern technology, the Zeppelin often served bizarre advertising campaigns. Capitalizing on the notion of the airship as a transport service, this milk-processing company suggested that its product would soon make its way to German colonies. Author's collection

and thriving on our lands."[142] The Zeppelin airships symbolized the economic and technological achievements of Germany.

Between 1908 and 1914, in the aftermath of the LZ 4 crash, the notion of sacrifice through commitment to technology came to occupy a special place in German society.[143] Count Zeppelin had poured everything into his obsession; in return, the German people gave him their savings. This ritualized giving would be repeated in 1912 for the creation of an air force, then several times more after World War I for various Zeppelin- and aviation-related causes. Not even the repeated crashes that airships suffered—albeit with no civilian loss of life—could shake public confidence in this technology. Count Zeppelin became the figurehead of this peculiar populism, and his machines the material embodiment of it.[144]

This form of the Zeppelin sublime was not monolithic. Rather, it embraced multiple cults that reflected the full complexity of the public's fascination with the old man and his machines. Count Zeppelin was not the revolutionary figure that popular reactions to his crash suggested he was. Far from challenging the symbols, parades, and rituals of the new German nation, he in fact helped reinforce them. His virtues were no longer merely individual, but societal. All good Germans were to emulate his work ethic, his demonstration of "loyalty and strength" in the struggle to travel through the "air sea."[145]

Like the various aeronautical events that informed the cultural pride of other nations, airships became part of the German cultural framework—a new symbol for an existing modernist culture. As such, they reconciled several notions of nationalism, some drawing on old traditions, others recently invented. The dirigible in Germany was in the unique position of depending on support from across the political spectrum and from all social strata, from the nation's children all the way up to the kaiser. This phenomenon reflects the emergence of a mass culture dependent on various technological marvels, the airship being one of many such, yet the airship had a strikingly universal and long-term impact. The "avant-garde" movement that supported the count prior to August 1908 included not just close associates but also politicians, professors, and industrialists, all of whom apparently shared the goal of creating an ideal that would suit everyone. Although they had hoped for a positive response to the count's airship activities, the magnitude of the response surprised them just as much as it did those who had ignored him. However shallow this response may have appeared to some, the airship was a material witness to the worth of the unified German nation sorting through contradictions and doubts about itself.

Critics have argued that mass culture reduces national identity to a substitute for lost community.[146] The case of the airship may seem to support this claim, but it also illustrates the liberating power of technology on the eve of World War I.[147] The traditional romantic protest against science and technology was directed less at the machine itself than at its uses. Germans accepted the airship's purpose: to achieve effective, controlled flight.

"The greatest of technological events for me and my contemporaries was the conquest of the air," reminisced Social Democratic leader Gustav Radbruch years later; "eyes filled with tears upon seeing a Zeppelin for the first time."[148] The airship offered an ideology of adventure tied to the glorification of its inventor, an ideology similar to the one once celebrated in knighthood rituals.[149] Just as Charles Lindbergh would describe two decades later a special link to his airplane, the *Spirit of St. Louis*,[150] so did Count Zeppelin earn sanctification through association with his machines. In becoming an integral part of cul-

Gone with the wind: the LZ 8 *Ersatz Deutschland,* intended to replace the crashed *Deutschland,* lies smashed against a hangar in Düsseldorf in May 1911 following a windstorm. Airships' vulnerability to the elements was a matter of serious concern to civilian and military authorities, yet the German public continued to support the idea of a great Zeppelin fleet. SHAA, B77/513

ture, the airship supplied a central element of German national identity in the modern age, much as the army and unification had done in the nineteenth century. Yet the enthusiasm and the spontaneous public reactions the Zeppelin evoked also confirm that a relative state of freedom did exist in imperial Germany,[151] contrary to the assertions that have been made about the state's authoritarianism in peacetime. Unfortunately, war would soon change this by sweeping aside notions of escapism and freedom in favor of militaristic nationalism. This would affect the image of the airship not only at home, but abroad.

CHAPTER 3

Zeppelin Myth and Reality
in the Great War

NOW RECONCILED TO INDUSTRIALIZATION, twentieth-century European society saw no contradiction between war and the technological sublime: the flying machine mirrored the virility of both the soldier and the nation he belonged to.[1] During World War I, Count Zeppelin's creations became the focus of militaristic pride on one side of the trenches and of deep hatred on the other. Both sides depicted the machine as things it was not, regardless of its technological performance. The military aspect of the airship antedates 1914, and popular visions of that aspect affected the machine's operation in this first total war. Propaganda played a substantial role in this process, masking failures on both sides while exaggerating meager successes. Mass culture absorbed the rhetoric, and sometimes even amplified it beyond what the censors intended. This sense of "sacred" warfare was so strong that it still informs the writing of military aeronautical histories as they focus on the usual mix of casualties and technical detail, ignoring the social lore that contributed to shaping the record. In fact, only a mosaic of experiences from the field to the home front can help us understand how the elevation of the airship into a German symbol of sublimity linked the social, military, and political dimensions of the Great War.

This mix of fantasies had surrounded the airship since long before the conflict, in the writings of nationalists and in the views of Count Zeppelin himself, who had always wished for his machines to function as weapons. Such attitudes provide the context in which military airship policies developed. While the prewar ideology of adventure in the air embodied a multitude of conflicting messages, by 1914 the colonization of the skies had taken on a clearly martial tone.[2] Popular and specialized warfare literature emphasized starker possibilities associated with the airship.[3] "The people of Europe must enter the third dimension empire under the leadership of Wotan," clamored one of countless nationalistic works.[4] Franco-German and Anglo-German antagonisms fueled one another and fed cases of mass paranoia.

Zeppelinitis

A series of airship scares between 1908 and 1914 affected public opinion in England in a way that in turn influenced the mood in Germany. In July 1908 Rudolf Martin granted the *Daily Mail* an interview in which he outlined plans for a German invasion of the United Kingdom: thirty-five hundred specially equipped airships would land 350,000 men in thirty minutes.[5] Impossible even by today's standards, the claim got people's attention. Several British experts responded by demanding that their government immediately establish a program to build aerial machines capable of countering the hypothetical German threat. Although nothing came of this demand, it had sown the seeds of panic. Lloyd George, a chancellor of the exchequer who later became prime minister, was in Germany at the time of the LZ 4 crash. He returned convinced that he had to do something to get his nation into step with German aerial progress.[6]

While the claims of German aerial armament at first evoked only a limited response from the British public, the constant coverage of the subject had a cumulative effect. Martin's designs for an invasion air fleet, presented to a Berlin audience in October 1908, again found their way into the British press.[7] Combined with a *Daily Telegraph* interview in which the kaiser had made bellicose statements, and other concerns about rising German power, such factors resulted in a major air scare in 1909.

The phantom airship scare of 1909 was part of the public hysteria linked to growing tensions between England and Germany. In early spring, reports began reaching British authorities and media about dirigibles flying near territories along the North Sea. Despite the factual impossibility of there being airships in these areas—neither British nor any continental machine had yet ventured over the sea—and despite the conclusion that several such sightings involved balloons or kites launched as hoaxes, by late May the rumor had spawned a full-fledged panic.[8]

Two witnesses twenty minutes apart reported having seen an airship. In one case, the witness claimed that the dirigible, partially deflated, was surrounded by crew speaking a foreign language. A journalist sent to check out the story reported finding newspaper clippings on the ground pertaining to aerial matters. The airship scare now focused on the threat of espionage, and many columnists warned that foreigners on apparently innocent sightseeing trips might be reconnaissance officers in disguise, updating maps. Reports of airship sightings continued throughout the summer, despite the embarrassment of British officials and the puzzlement of the German media, which began questioning the sanity of the English.[9] By June, however, with no further confirmation of the sightings, the panic died down and was replaced by a general interest in technical advances and the creation of an air fleet.

However, the fears resurfaced four years later. In October 1912, rumors began circulating anew that a German airship had crossed the North Sea and flown over the garrison town of Sheerness, near the Thames estuary. Although the claim, founded on officers' reports, proved to be based on nothing more than an unidentified noise, it refocused British attention on the matter of air defense against airships. Despite strong denials from the German side, "Zeppelinitis" began to spread again.[10]

Zeppelinitis also afflicted British military circles. Flooded with media stories of airship visits to the British coast, the Admiralty rushed to establish plans for an effective air defense. Concluding that no airplane could overtake an airship at high altitude, strategists recommended the establishment of an artillery-based defense. Renewed parliamentary debate ensued, prompting the politician and newspaper magnate Lord Northcliff, concerned by the lack of concerted political action, to ask H. G. Wells to write an opinion. Wells published three articles in April 1914 on the future of the war in the air, arguing that new technologies, rather than established strategies, would determine the winner.[11] Politically, these articles had little impact, as the government and the opposition continued to argue about the proper policy with respect to aeronautics and, in the greater scheme of things, toward Germany. The direct impact on British air policy remained limited, even though the media continued to speculate on the dynamics of air war.

Press accounts fed tensions by focusing on fantastic war technologies. Descriptions of stealthy, silent airships accompanied illustrations of machines equipped with death rays capable of building an air bridge. *Pearson's Magazine* noted that the projected capacities of aircraft in war were quite sobering, adding, "These giant airships will carry within themselves sufficient power to enable them to make headway against a storm of tempest-like strength."[12] To feed its coverage, the British press relied on German publications that reported on rigid- and nonrigid-airship activities, and claimed an "airship gap." Parliamentary debates ensued, forcing a young minister, Winston Churchill, to acknowledge that Britain had no equivalent to Germany's rigids. This was more than a matter of military superiority, however. The honor of Great Britain hung in the balance: "We, the officially supreme nation of the earth, the arbiter of world policy, are unable to return a call made on us by a powerful rival," commented one technical expert, rather melodramatically.[13] An unmistakable expression of British concern was heard in London's Drury Lane Theater. There, a spy drama, *Sealed Orders,* involved a British patrol boat shooting down an unidentified airship.[14]

British Zeppelinitis also infected Germany. The German media often printed translations of articles from their British counterparts, informing Germans of how great and worrisome their country's dirigibles were.[15] One flabbergasted German magazine columnist noted that "never has a German airship overflown England's green turf . . . yet we recently witnessed an entire people panicking at Zeppelins."[16]

While the British had yet to see a German airship firsthand, the French had already had a chance to examine one closely. On 3 April 1913 the army Zeppelin LZ 16, flying to the Baden-Oos airship base, encountered serious trouble. Five hours into the flight, the crew lost their bearings under heavy cloud cover. The engines failed, and the venting of hydrogen to maintain altitude against heavy lifting winds seriously depleted the machine's reserves. Finally the airship broke cloud cover, and the crew recognized French flags fluttering atop buildings; they had reached Lunéville. The captain ordered the dumping of all ballast, hoping to make it back to Germany, but to no avail. An emergency landing followed in a clearing that turned out to be an army training grounds filled with cavalry units undergoing inspection. The crew disembarked, and French troops began examining the machine.

French authorities gave the crew a courteous welcome, and diplomatic channels cleared the way for civilian repairmen to cross into France. The airship departed the following day under their control, and cars dropped the officers off at the border.[17] While the crew later confirmed that they had been treated kindly by the French officers, they also noted that the officers had carefully photographed and measured the airship. Furthermore, the population of Lunéville had displayed considerable animosity toward their unexpected guests, to the point that the French officers had felt compelled to provide the uniformed Germans with civilian clothes. Local people had approached the soldiers who were guarding the machine and urged them to burn it, and some of the German mechanics had been stoned.[18]

Newspapers in both nations embellished the kernels of truth to the point that it became unclear just how nasty the local population's behavior had been and what damage, if any, the incident had done to German national security. German reports on the French interest in the machine confirmed the German public's belief that Zeppelin's machines were wonder weapons that would ensure Germany's dominance of the geopolitical heartland.[19] The Lunéville incident thus confirmed, as the Saverne affair would do a few months later, that the nationalist revival that had gripped both France and Germany at the turn of the decade was now in full swing, fueling popular reactions that were bound to surprise leaders on both sides of the border.[20]

Lunéville and the spread of British Zeppelinitis also illustrate that fears from abroad were feeding a German sense of encirclement—a feeling that any foreign call for a military buildup might spell doom for the homeland.[21] One had to respond to such threats. The buildup to an arms race that included dirigibles, however, started very slowly, taking years to spread from rumors and potboilers into war cabinets.

Technology and War Planning

As Count Zeppelin discovered soon after the crash of LZ 4, the path to military orders was paved with skepticism, enthusiasm, and politics. The national outpouring of sympathy that had occurred in 1908 included expressions of support for a future in which the airship would play a military role. The count capitalized on this mood, hoping to force the military's hand in the creation of an all-Zeppelin fleet.[22] It did not happen. Although the old man had sought from the start to attract military attention to his experiments, the response had been lukewarm at best.[23] Zeppelin thereupon shifted gears and readily accepted the "cultural mission" associated with his invention, though he continued to predict that, once perfected, his machines would have multiple military applications.[24] To the letter-writing Luddite who once castigated him for building machines that might destroy cities, he responded by scribbling in the margin, "No answer to this special saint."[25]

Meanwhile, the military continued to drag its feet. Several German government officials as well as military officers had published pamphlets and studies purporting to demonstrate the military value of the airship despite its current weaknesses. Observation of troop movements from the air, although practiced since the French revolutionary wars, might improve thanks to an airship. If fitted with newly perfected radios, it could relay enemy artillery and troop positions to military headquarters.[26] Several experts were already warning that the airplane might be better suited for such missions because it was more difficult to spot from the ground, and its size made it less vulnerable (speed was not yet a factor), yet they acknowledged the capacity of the airship to remain in the air longer.[27]

In 1909 the German army conducted special air maneuvers near Cologne using all three types of airships. While the general purpose was to devise a set of protocols for the effective use of new technologies, the specific goal was to determine which type was fastest, most maneuverable, and best suited for "the duties of war." The results were inconclusive—all three types suffered from the usual wind-vulnerability problems—but the official report was noncommittal, on grounds of military secrecy.[28] The loss of a military airship during similar maneuvers the following year was blamed by the count's associates on the army's handling, in an effort to preserve the Zeppelin reputation. The claim did little to boost military orders; the army wanted no new Zeppelins. The fact that its maneuvers were conducted near the French border, however, sent a clear message that this aeronautical technology would play a role in the next conflict, even if the specifics remained unclear.

Despite these disappointing results, populist nationalism continued to feed on publications that made fantastic claims for airships. The writings of Rudolf Martin were typical. While stressing culture, Martin's works also addressed more nationalistic and confrontational themes. One book called for friendship

The military-industrial complex. As World War I shifted to a total war, the Continental rubber company changed its ads to show its role in helping fight the conflict. A crewman atop a humanized observation balloon watches enemy lines for General Hindenburg while airplanes and an airship fly around it. ASTRA with England but imaged such rapprochement as depending upon the existence of a strong German air fleet, which would weaken Albion and open it and its overseas territories to possible invasion.[29] In effect, the airship offered a chance at restarting the scramble for colonies. Similarly, other writers of pulp fiction such as the popular Paul Scheerbart emphasized how useless established weaponry would be in the face of the new airship fleets that would be able to launch an attack and destroy an enemy in a matter of hours.[30]

Books provided aviation nationalists with entertainment, but to voice their own views, they joined aeronautical leagues similar in structure to military ones. The army and navy leagues represented a peculiar military-industrial way to manipulate and channel public pressure on the government and thus increase the size and budget of the services. Taking his cue from the success of such organizations, Karl Lanz, co-founder of the Schütte-Lanz airship company, created a German Airship League, which issued calls for aeronautical buildups in the army and the navy. However, the new association failed to increase its membership after the first wave of enthusiasm.[31] Further hindrances came from the military. While the army was sympathetic to Lanz himself—it hoped that competition would prompt the Zeppelin company to improve its designs—the navy feared that public pressure would force it to scale down its ship budget in order to support unproven dirigible designs.[32] Furthermore, the Airship League and other such aeronautical associations rarely had a single, exclusively military purpose—some focused, for example, on calisthenics—nor did they federate themselves under a single banner. With popular pressure thus dispersed, only members of the government sympathetic to Count Zeppelin might effectively push for the purchase of military dirigibles.[33]

In Reichstag debates, calls went out for more airship subsidies—handouts along the same lines as those the nascent automobile industry was receiving. That way, some felt, the imperial army would have a choice of already advanced models should war break out. A few Socialists challenged this utopian view of war in the air, calling it a glorified cockfight.[34]

Despite public and parliamentary enthusiasm, the German High Command remained skeptical about the military benefits of the airship that would purportedly come about once minor defects were ironed out (such as sensitivity to wind). From a projected fifteen Zeppelins for the army in 1908, Chief of Staff Helmuth von Moltke revised the number down to three by 1911, partly in view of recent airship mishaps and partly because of the need to support airplane development on the same budget. Yet by late 1912 the international situation had prompted a new arms race and the drawing up of new military plans.[35]

The variable in the military equation came from France. Count Zeppelin had used the argument of airship competition with the French as early as 1906 to press for continued governmental support. Within a year both nations had declined, de facto, to follow the renewed provisions of the 1899 Hague Peace Conference, which had sought to forbid the throwing of bombs from flying vehicles.[36] In 1909, even as Louis Blériot was crossing the Channel, the French corps of engineers, then in charge of evaluating airplanes as potential machines of war, initially rejected them in favor of building dirigibles, possibly after seeing the German public's enthusiastic reaction to Count Zeppelin.[37] France and Germany thus found themselves involved in an airship competition of sorts.

The first aeronautical expression of renewed tensions between the two countries was a drive in Germany to increase funding for military aviation through a public fund-raiser. The movement was led by Prince Heinrich, brother of the kaiser, together with other famous personalities.[38] While the reasons for supporting military aviation at this time were clearly military and political, the arguments were put to the German people in a way that appealed to their sense of the uniqueness of the Zeppelin experience. Cool-headed Germans did not just jump into things the way "Latin lands" did, explained government-printed leaflets, but rather deferred action until the matter in question had become serious and crucial to the honor of the nation. That time had come; now was the time to repeat the experience of 1908 and rescue the beleaguered airplane industry.[39] Echterdingen and the return of Count Zeppelin had proven what popular support could accomplish, and now airplane development would follow the same path, "for the people and by the people."[40]

The call to air arms was being heard in several European nations at about this same time. The difference in Germany was that business and governmental supporters of the airplane, despite a call to arms dating back to the 1909 visit of French and American aviators to Germany,[41] had lost out to airship enthusiasm. They now desperately needed to boost public confidence in airplanes, yet they were relying on the Zeppelin mystique to do it.[42] The German government, which four years earlier had feared the popular reaction from all across the political spectrum in support of an old man and his invention, now saw that phenomenon as a way to channel public support toward another project. The multiple perceptions of the count's success were brushed aside in favor of the one that counted the most: *raison d'état*. Any airship success achieved during the time of the fund drive barely received mention, to ensure that no one would protest that funding for airplanes was a misallocation.[43] As the Bavarian representative in Berlin, Count Lerchenfeld, put it to his governor, "I am personally not an aviation enthusiast, but the thing is in fashion."[44] The Zeppelin symbol thus proved its adaptability once more, this time working in favor of the airplane.

Yet dirigibles were gaining ground in the military, too. The second expression of Germany's toughened aeronautical military stance came, to everyone's surprise, in modified war plans for airships. In 1912 von Moltke once again changed course and announced that the army would need no fewer than twenty airships within three years. The general staff, now working on a modified Schlieffen attack plan (which called for the defeat of France within weeks following an invasion through Belgium), felt that the airship might provide the necessary strategic advantage. In briefings, however, this view was framed in the context of Count Zeppelin's favorite argument: France had a flotilla of observation dirigibles, and therefore so should Germany.[45] While this kind of argument was common in civilian nationalist circles, it was new to the military, where concerns over practicality and the need to develop airplanes had so far been more important. Illustrating the reversal, von Moltke claimed that the Zeppelin weapon could have a "practical and psychological impact" during the opening moves of a war.[46] The army concluded that it need not use the Zeppelin company's machines; while Zeppelin had made great technical strides, the superiority of its machines had perhaps been overstated in some of the early reports. Further documentation implied that machines built by either Zeppelin or the other German rigid-airship company, Schütte-Lanz, would eventually serve as tactical tools, airlifting troops to the front and evacuating the wounded to field hospitals.

Debates in the Reichstag over the 1913 budget had raised the issue of such potential, prompting the French and Russian military attachés to express their concern.[47] After all, the airship, serving since 1910 as an instrument in "peaceful competition of the air,"[48] had now become part of the military arsenal. Talk of Zeppelin airship sales abroad, an idea floated earlier as a means of garnering governmental support for the rigid airship, had ceased in the name of protecting trade secrets and national prestige: to sell off such treasures would suggest that Germany saw no value in them. Even Austria-Hungary could not possibly receive the most recent airships. Any sales negotiations would require prior approval from the War Ministry.[49] Reports spread that airship production was reaching one a month in Friedrichshafen, giving the impression that German air power was on the rise.[50] In France, where nerves were still raw from the Lunéville incident, popular reactions to the airship seemed to confirm this: "They cannot swallow our machines," stated Zeppelin associate Fraunhofen following a visit to Paris in the fall of 1913.[51]

Within a year, Germany had increased its military airship force to twelve machines and twenty-six airship fields. The DELAG airship transport company, formed initially as a temporary solution to the dearth of military contracts, served the military well. As part of its campaign to increase interest in airship travel, DELAG often offered free rides to the German elite, including many officers. Consequently, of the thirty-four thousand passengers DELAG carried up to World War I, some twenty-three thousand never paid for a ticket. Costs were apparently covered through a combination of subsidies from cities and income from regular passengers. The officers who traveled under this arrangement would go on to serve aboard airships in wartime, having benefited from the informal civilian training.[52] Count Zeppelin's constant effort to paint his invention as a war machine had convinced many observers.[53] Colonel Erich Ludendorff, for example, congratulated the count on his seventy-fifth birthday and stated that he looked forward to the machines' performance in the next conflict.[54]

Ludendorff's optimism ignored the many problems that still plagued airship technology. Although making good progress, airship technology was not mature. The haphazard way in which aeronautical design and construction had proceeded until World War I helps explain this state of affairs.

Despite changes in the operating structure of the factory that brought it closer to the scientific method, the Zeppelin design team of engineer Ludwig Dürr took a conservative approach to innovation. The reasoning behind the slow progress included the concern that for every increase in the diameter of an airship's cross-section, new loads, tensions, and pressures were created, all of which required extensive calculations. Until new engines could be designed and installed, there was no point in running the risk of compromising an established design. For example, the company could have switched to using duralumin soon after the invention of this aluminum alloy in 1909; it had the same properties as the original metal but was lighter and sturdier. Nonetheless, engineers avoided using it until 1914, apparently out of dissatisfaction with the way the alloy behaved during component machining.[55] This conservatism also masked Count Zeppelin's personal pride—his conviction that he alone had arrived at the proper solution to the problems of airship design. Several of the rejected suggestions regarding aerodynamic improvements came from engineers who had indirectly criticized his early designs. The count's personal pride as well as his fame thus shaped the engineering culture of his company.[56]

Until 1912, all Zeppelin airships looked like crayons. Zeppelin himself argued that this shape was the easiest to build and that the various alternatives that had been suggested, including a more pronouncedly spheroid shape, would not offer substantial improvements in performance. Now, however, two factors forced a change of attitude: the competing airship firm Schütte-Lanz, and the German navy.

By the time war broke out, the Schütte-Lanz company had patented several new rigid-airship designs. Launched with the tacit support of the German army, which hoped to see competition develop with Zeppelin, Schütte-Lanz had introduced several notable improvements to rigid-airship design. New aerodynamic shaping took into account calculations done in a wind tunnel at the University of Göttingen, while the internal structure was helicoid (a system of crossed spirals) and made of wood rather than aluminum. Although wood may seem to have been an outdated choice in view of aluminum's inherent solidity, the tensile strength of various woods was well known thanks to shipbuilding experience. Several airship projects were premised on the use of wood, but only that of Schütte-Lanz became a reality.[57] The rationale for the use of wood was the belief that it would allow for effective radio communication with the ground without any risk of explosion due to a buildup of static electricity. This speculation turned out to be unfounded, but it nonetheless prompted the building of all Schütte-Lanz machines from wood.[58]

Schütte-Lanz offered other improvements, such as enclosed control cars and engines placed one behind the other rather than side by side. This new arrangement ensured that if one engine broke down, the other would compensate by propelling the dirigible on its axis rather than sideways, as had happened in earlier Zeppelins. Machine guns were also mounted on the top and the rear of the airship.

Several engineers in Count Zeppelin's organization pushed for the adoption of similar improvements. The ultimate thrust for change, however, came from the German navy. Early in its relationship with the Zeppelin company, the navy had pressed for bigger airships but met with resistance. However, the dearth of airship orders from the military ultimately forced the company to agree to a series of changes, some of which were improvements originally patented by

One of the improvements to military airships involved the installation of rear-facing machine guns to protect Zeppelins from enemy fighters. Based on the Schütte-Lanz defense system, the installation had limited effectiveness and was eventually removed from later models. ASTRA

Schütte-Lanz. By this time Schütte-Lanz had ceded its patent rights to the German War Ministry in exchange for several contracts. The Zeppelin company thus now had access to these improvements.

This does not mean that the Zeppelin company completely ignored the suggestions of its own young engineers. Indigenous improvements included the redesign of rudder and elevator controls, placing directional boards behind the vertical stabilizers to ensure a better turn ratio (before, the directional boards had been attached to the horizontal stabilizers).[59] But designs continued to evolve at a very slow pace, in part because of a lack of installations capable of handling bigger designs. Not until December 1914, with its twenty-sixth airship, would Zeppelin begin producing bigger dirigibles that clearly surpassed the performance of earlier machines and qualified as full-scale weapons.

While war may accelerate technological development, it also accentuates any invention's flaws.[60] Before the war, supporters of the airship had claimed that once the technical problems were ironed out, the machine would offer the perfect combination of speed, range, and payload.[61] As we shall see, Zeppelins improved in many of these capacities yet failed in their task as bombers while proving useful as reconnaissance machines.

The first Zeppelin to incorporate substantial improvements was LZ 26, completed in late 1914. Its wider diameter gave it greater volume and, therefore, greater lifting power. The control cars, initially open, were rebuilt to Schütte-Lanz patent specifications and enclosed. The galley that allowed crew to move between them was placed inside the hull rather than below it, thus improving aerodynamics and access to the ballonets. LZ 26 led the way for further technical progress, this time thanks to the work of Zeppelin designers.

Among the engineers in Friedrichshafen, Paul Jaray deserves special mention, for the pattern of his suggestions makes clear the conservative approach that prevailed at Zeppelin and the resulting tensions that arose between generations of engineers. Born in Vienna, Jaray joined the Zeppelin company in 1914 at age twenty-five, charged with the task of devising better aerodynamic shapes. Although the company had benefited from a number of Schütte-Lanz patents, these were based on helicoid wood structures. To achieve greater speeds, the company needed improvements based on aluminum and designed for use in bigger machines. By 1915, extrapolating information from the work of Johann Schütte and of the Frenchman Charles Renard, Jaray devised ways to calculate the optimum airship form based on diameter cross-section, volume, and stress points.[62] In layman's terms, he proved that airships did not (as was commonly believed) require a tubelike cylinder in the middle for effective steering, and that an increase in payload was better achieved by an increase in diameter than by an increase in length. Furthermore, as an unwavering advocate of streamlining, he called for redesigned control cars and propellers but ran into opposition.

Jaray promised a substantial increase in speed when the first ship of a new 1915 series was tested. Eyebrows were raised at his projection of a cruising speed of over 95 kph—unheard of for an airship of this time—and indeed, the best result actually achieved was only 88 kph. The military pilots dismissed Jaray as a crank. But then he learned that the engineering supervisors had ignored his recommendation that redesigned propellers be installed on board the series. He protested vigorously, and when the mistake was corrected, his projections were not only confirmed but exceeded.[63] Anecdotal though it may be, this episode illustrates both the conservatism of the Zeppelin approach to design as well as the role the military played in the building of new airships.

Jaray never saw his ideal streamlined hull implemented in wartime, yet some of his calculations did help build bigger war machines. As the war dragged on, it was increasingly thought that more powerful technology might break the stalemate. The notion that bigger was better spread among Zeppelin crew, too. It was joked that a 1916 Schütte-Lanz machine of 36,000 cubic meters "still needed to grow in its shed" to catch up to the bigger Zeppelins.[64] Joking aside, the 1916 Zeppelins of 56,000 cubic meters were indeed impressive new "height climbers" that signaled great progress.

The new wave of "war-class" Zeppelins benefited from innovative engineering ideas and feedback from the navy. The navy ordered that the projected six engines be cut to five and that fuel autonomy be reduced to thirty hours' worth; it also requested lighter hull girders, no crew quarters, and a smaller control car. Further improvements included the reduction of skin friction. The upper part of the hull fabric was usually left porous to allow for the escape of hydrogen seeping out of the gasbags; this diminished the risk of explosion from the gas remaining in suspension. Now, that section was doped to diminish air resistance. As for the gasbags, initially made of three layers of goldbeater's skin, they were eventually replaced with lighter silk fabric.[65]

The end result was remarkable: each new machine produced in 1917 boasted further improvements, reducing weight while extending operating distance from 7,600 kilometers to over 12,000 kilometers. French intelligence analysts were amazed at the progress that had been made in just one year.[66] Their

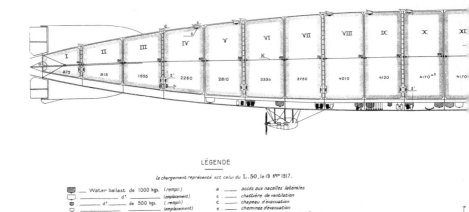

LÉGENDE

Le chargement représenté est celui du L.50, le 19 8ᵉʳ 1917.

____ Water ballast de 1000 kgs. *(rempli)*
_____ dᵗ _____ *(emplacement)*
_____ dᵗ ____ de 500 kgs. *(rempli)*
_____ *(emplacement)*
____ reservoir à essence de 400 litres(rempli)
_____ dᵗ _____ *(emplacement)*
_____ hamacs
____ bombes de 300 kgs.
_____ dᵗ 100 kgs.
_____ , dᵗ 50 kgs.

a ____ accès aux nacelles latérales
c . ____ chattière de ventilation
C . ____ chapeau d'évacuation
e . ____ cheminée d'évacuation
E . ____ accès à la plateforme de mitrailleuse
K . ____ câble
m . ____ plateforme de mitrailleuse
n . ____ pantalons d'eau
S . ____ soupape commandée
S'. ____ dᵗ automatique.

report on Zeppelin L 49, which had crash-landed almost intact in Bourbonne-les-Bains in October 1917, sang the machine's praises. L 49 was some 6 metric tons lighter than its predecessors, which gave it a higher cruising altitude and greater bomb-carrying capacity. So impressed were the Allies that L 49 became the basis for a case of reverse-engineering (making plans from a built machine) that would lead to the construction of the American *Shenandoah* airship after the war.

However, even the improved airships continued to display weaknesses. Engine power remained a central concern, since advances in propulsion technology had not kept up with the quickened wartime pace of developments in other areas. Simply put, 1916 airships used 1915 engines designed for 1914 projected combat conditions. At high altitude, not only might coolant water freeze, but the engines might either quit or slow the airships from 100 kph down to 70 kph. The problem would finally be solved through the introduction of altitude-proof, long-range engines in December 1917.[67]

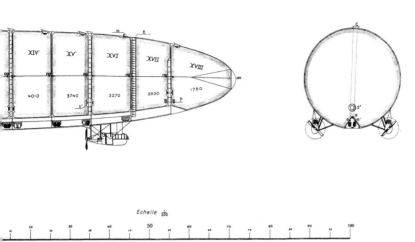

Echelle

20   30   40   50   60   70   80   90   100

A schematic profile of the navy Zeppelin L 49, which was brought down over France. Engineers took the airship apart and developed a series of blueprints and reports to evaluate design improvements since 1914. The plans were eventually passed on to France's allies and served as the basis for the design of airships after the war. *Extrait de l'étude sur le Zeppelin L 49 (1917)*, via SHAA

Improvements like this strengthened the airship's ultimate trump card: autonomy. The airplane, despite amazing successes (the first long-range bombers to attack London were designed in 1917), remained weak in this area, and supporters of the airship were quick to emphasize this fact. Some went so far as to recommend that airships be used as part of a long-range supply system. An experiment to that effect took place in November 1917, when a specially fitted machine flew from Bulgaria to East Africa to deliver supplies and demonstrate support for the defenders of the German colony there. Although L 59 (also known as LZ 104) turned back before completing its mission, the distance and duration of the flight (some 5,700 kilometers over 95 hours) were an impressive feat, and one that was never repeated.

Technology thus continued to shape the symbolic power of the airship in wartime, disappointing some expectations of its performance and unexpectedly fulfilling others. While the supply-ship idea, less glamorous than that of a flying attack-cruiser, was tried but once, the African expedition set the stage for the airship myth-making that would occur after the war, when the experiment would be cited as proof of the airship's long-range capacity.[68]

Demonizing the Dirigible

Where engineering failed to offer bona fide improvements to the airship, rumor and propaganda stepped in to fill the gap. Thanks to these efforts, the airship acquired, very early in the war, a reputation quite disproportionate to its actual effectiveness.

In 1914 the presence of airships on the battlefront was a great novelty, and the machines were automatically associated with Germany. French troops once shot down one of their own observation semirigids on the assumption that it was a "Boche," while in England, antiaircraft gun defenses were ordered to fire only at airships, since they were sure to be German; any airplanes might be British patrollers.[69] Such visions of airship doom began to affect the home front in the first year of the conflict, before any bombings had occurred.

World War I was fought not only with new industrial weaponry (like the Zeppelins) but also through a plethora of popular arts: posters, postcards, and so on. The airship figured prominently in such materials, German and Allied alike. Allied depictions painted the airship as a tool of evil accompanying the Teutonic barbarian on his quest for domination. Sometimes these depictions showed the airship doing its dirty work alone, without a human accomplice, leaving in its wake a trail of civilian victims: this enemy was especially monstrous, because technology had rendered it faceless. Pictures like these relied on emotion to get the point across. An airship silhouette looming over a female figure lying on the ground delivered a clear message: Men must rise up against such barbarity. This was an enemy undeserving of respect, for it had, with complete indifference, struck down a helpless civilian. To prove one's honor and not sink to the coward's level, there was but one thing to do: enlist and defend the homeland.[70]

Where posters and drawings left off, the caricature picked up. Germans were time and again represented as pirates or as monsters with bestial features. Such caricatures were designed with both the home front and the trench soldier in mind. Funny, they could entertain during the hours of boredom; angry, they might incite the kind of rage that some felt was needed to break the stalemate.[71] In traditional propaganda, political-military themes (such as battles) were typically kept separate from cultural-warfare themes (negative national

Inspiration to victory. The airship became a favorite motif of propaganda both on the front and at home. "High in the air, fresh German courage rises invincible," proclaims the motto. An infantryman sent this card to his cousin in Bavaria in June 1917. By then, Zeppelin losses had reached alarming numbers. Author's collection stereotypes). Allied propaganda against Germany, however, tended to mix these themes together, in such symbols as the kaiser himself and the airship. This mix was especially evident in postcards.

Postcards made prolific use of the airship theme. The medium was a double-edged sword: on one side appeared the picture; on the other, news from home or the front might inform its recipient on unrelated topics.[72] The message on the picture side suggested the mood associated with the subject, often a potent mix of anger and humor. While propaganda obviously sought to manipulate public opinion, it would not have succeeded without an underlying receptivity to the ideas it was trying to convey. The private publishers who often produced these war postcards sought to capitalize on popular patriotic culture while in turn contributing to it. In France alone, some twenty thousand different war-card designs appeared (and circulated), all emphasizing the stupidity and cowardice of the enemy.[73]

The French viewed Count Zeppelin's machines as the quintessential instrument of a war-mongering kaiser. Humorous representations from early in the war showed, for example, an airship trying to tow away major Parisian monuments. After the first airship bombings, of course, things changed, but not necessarily in the way the Germans had hoped. The effects of the bombings played right into the hands of Allied propagandists. Censor-approved photographs depicting survivors and damaged buildings heightened the popular mood against the airship through a best-selling series entitled "Horrid Crimes of the Boche Pirates."

Reportedly more effective than the simple reprinted photograph because of its entertainment value, the caricature postcard representing the airship often sank to the level of the ethnic joke.[74] From a Prussian-helmeted pig surrounded by airship flies, to an unhappy Teuton (usually the kaiser or Count Zeppelin himself) bemoaning his lost bratwurst, these drawings emphasized national stereotypes and even racist ones (for example, equating the German soldier with an African cannibal), all in the interest of diminishing the human dimension of the enemy. In Russia, where a good one-third of all postcards caricatured the kaiser, a variety of inflatable objects, from bubbles to airships, surrounded him, all of which ended up deflated, providing an apt metaphor for overblown Germanic aspirations.[75]

Another central theme of home-front culture was childhood. As part of the buildup to total war, the adult world devised books, toys, and school lessons of a warlike slant to ensure a moral and intellectual mobilization of children.[76] The airship figured prominently in this effort as well. Christmas-tree ornaments in Berlin, board games in Paris, "anti-Zep" candlesticks in London, and books on all sides used the airship as a focal point of child war culture.

One theme of this material was that children were as excited about airships as adults were. The scenes painted by the cartoonist Francisque Poulbot, of the Parisian *Journal*, are typical: "The tired, irate neighbor: 'So, every night, no sleeping allowed?' The mother of the screaming boy: 'Why don't *you* try and get the Zeppelins to show up? He wants to see them.'" There was a fine line between smiling at a child's unawareness of danger and representing the child as a hero, but adults happily crossed it. A French play presented the exploits of a young girl flying against an airship, and postcards often took to representing children in adult roles.[77] In Germany, a young boy posed with a balloon for a card entitled "The Little Airshipman," while in a French counterpart, a child smoking his father's pipe dares the ugly airship to scare him. By placing children in such contexts and portraying them as "pre-adults" braving danger, adult fantasies were acted out and wrath against the enemy became complete.[78] Portrayals of children in "soldier games" sometimes also played a role in pro-natalist policies—declining demographics were a central concern in France—and were considered an essential element of support for front-line men far from their families.[79]

Throughout such representations, the goals of the war were never explained: Zeppelin attacks were depicted as ends in themselves that demonstrated either Germany's superiority or its utter barbarity.[80] Zeppelins, the "Big Bertha" cannon, and other methods of bombardment made for new representations of Thor's hammer, striking randomly. Some French youth publications quoted with glee the German poet Heinrich Heine's 1834 prediction that the Germanic god would rise up to destroy cathedrals:[81] paganism reincarnated as technology was the ultimate proof of the Teutonic hell.

The relationship of the Zeppelin to popular perceptions of the war had two sides. In Germany the airship evoked hopes of breaking the stalemate of the trenches. Among the Allies, it rallied the public to the cause of the war.

The best Allied anti-Zeppelin propaganda, obviously, came out of the shooting down of an airship. Such an event offered an opportunity both to honor new heroes—the pilot or ground unit credited with the deed—and to reassure the edgy population that something was being done to stop the bombings.[82] The machine's carcass also became fodder for the media, nearby inhabitants, and of course the military. In hopes of learning more about how airships were built,

Un loustic qui n'en a pas peur

A kid not afraid. World War I–era popular arts made abundant use of the child motif to boost natalist policies. Here a future sailor smoking Dad's pipe looks placidly at the Zeppelin rumbling overhead. Approved by the censorship board, this postcard traveled during the early part of the war, and its sender inscribed on the back, "Long live the class of '35." Author's collection

Zeppelin wreckage as a war trophy in Saloniki. While flying a reconnaissance mission over northern Greece on 5 May 1916, the army airship LZ 85 was brought down by cannon fire. Twelve men survived and were taken prisoner. General Sarrail, commander of the French forces defending Saloniki, heard his troops cheering as the airship crashed and ordered the hulk to be preserved as a source of encouragement to his men. Postcards depicting the wreckage also circulated in France and Greece. SHAA, B89/1380

military intelligence services spent days trying to recover airship L 15, which had crashed in the Thames estuary in April 1916.[83] On land, the fuselage of L 33, which crashed in September 1916 in the English countryside, remained propped up on poles for months so that engineers could study it.[84] LZ 85 met a similar fate in Saloniki, the hulk being kept for a while as an inspiration to the French and Greek troops defending the city, before being shipped to Britain for study.[85] Even reduced to a metal skeleton, the airship's sublime proportions continued to fascinate onlookers.

Surviving crew members remained, of course, the most valuable prize to intelligence services. An unintended consequence of the survival of any crew members was to impress upon witnesses that those manning the monsters were just as human as the people below. In most cases, though, a downing resulted only in charred bodies and molten metal. French officers inspecting the crash site of LZ 77 had to content themselves with an aerial view of an airship base in the pocket of a mechanic's coat; the singed letters he carried were from home, not from the High Command.[86] Their British counterparts, however, once had the luck to find a usable code book. Luck returned to the French following Germany's attempt at hitting London with eleven airships on 19 October 1917: four went down, including L 49, which crash-landed but whose crew did not have time to set the machine on fire.[87] A photographer was present to immortalize the event, and a kind of carnival atmosphere took over the area, as pilots flew from their bases and hundreds of civilians came to gaze at the 134-meter-long monster: "What a monument," recalled a Parisian witness; "imagine, this is about as long as the distance between the Madeleine church and the Parliament."[88] The ship was eventually taken apart, most pieces being shipped to the army museum in Paris for display to its embattled inhabitants, while for years the mantlepieces of farmhouses in the area of the crash bore witness to the event with "austere rectangles of waterproof dope surrounded by clippings of the war."[89]

Whereas civilians in previous wars had usually been mere spectators and their involvement incidental, this time they became potential targets; France and especially England experienced the first of the twentieth-century strategic air bombings. The first raid on London occurred on 19 January 1915. Until then the British media had taken care to minimize the number of "airship scare" articles; the press dismissed the Zeppelins as big toys, to the point that during the first raids, some civilians were so unconcerned that they refused to darken their windows.[90] England had long taken pride in its "splendid isolation"; the Wright brothers and Louis Blériot had hardly shaken that pride, and even the prewar airship scares had made little real impact on British confidence (a few defense contingency plans appeared but were not implemented).[91] Following the first raid, however, the *Times* of London remarked that the heart of the British Empire was no longer safe from "the sight of a foe and the sound of an enemy missile."[92]

Although people were generally frightened by airship raids, many witnesses admitted to being strangely attracted by the sheer display of power, leaning out of their windows to see better instead of running down to the basement. As the Reverend Andrew Clark recalled:

A Zeppelin model in a British hospital, ca. 1916. In the wake of the airship raids on London, airships came to be viewed as the very personification of the hated "Hun"—a kind of reverse sense of the Zeppelin sublime. Placed in the clinic's recovery ward as a conversation piece, this cardboard and tissue-paper model is "attacked" by British fighters suspended above it. SHAA, B77/531

> Miss Eliza Vaughan's (the suffragette, of Rayne) great ambition has been to see a Zeppelin. Hearing that Zeppelins were expected on Su. night (1 Oct.) she went out, with a bag of peppermint bulls-eyes, at 9.30 p.m. and sat on the bridge at Rayne in the midst of all the roughs of the place. Just before midnight, when her mouth was sore with sucking peppermints, and she was very bored with the conversation of the yokels, she managed to see the Zeppelin.[93]

Once information about victims began to circulate, however, the public's curiosity turned to anger. When an airship was downed, although some people flocked to the site for souvenirs,[94] many others were angry enough to run to the

A war allegory of 1914: the Triple Entente's airplane downs the Teutonic Zeppelin. This artist's conception reflects the naive optimism that characterized the opening months of World War I. The Englishman says, "Here I am!" The Frenchman adds, "I am ready." The Russian chimes in, "Me too." To which the German responds, "If all three show up, I'll burst." In reality, it was months before an airplane managed to hit an airship. ASTRA nearest enlistment office and sign up, unaware of the hell that awaited them. Ironically, as the writer Philip Gibbs recalled, soldiers who went on leave discovered a home front unaware of their suffering and steeped in myths of glorious warfare. Feeling estranged from their fellow citizens, some trench veterans actually came to wish for more Zeppelin raids, to give civilians a small taste of how the face of war had changed.[95]

Some fifty-seven airship raids hit England, causing almost two thousand casualties and inaugurating the need for protection from strategic bombing. But fighting a Zeppelin in the air was unlike anything previously attempted. Prewar plans had either foundered on political gridlock or proved ineffective, and pilots had yet to perfect antiairship tactics. A formal air-intrusion warning system would not be in place in Britain until 1916.

As a new weapon, aviation was still an experiment, its limits constantly evolving. Artillery projectors searched the skies and sometimes found an airship, but when artillery shots failed to find their aerial targets, airplanes would take off in the hopes of tracking down airships hiding behind cloud cover. Even when the search was successful, the attack would often fail. Early in the war, airplane gunfire had little effect on airships, either because the mammoth machines flew too high or because airplane weaponry did not yet include incendiary bullets.[96] Pilots could certainly not be called upon to crash their planes into a Zeppelin for lack of proper weaponry, even though, as one commentator noted in his diary in 1915, this was being done on the ground every time soldiers went over the top.[97]

As in London, not everyone in Paris was angry at airships. The first Zeppelin raids, whether real or imagined (false alarms were frequent), became a kind of spectacle, and many people were irritated by the authorities' recommendations that they stay indoors and turn out the lights: "Why dress Paris in black? An airship will always find the city, even if only by the blare of stations and suburban factories. Is it, then, to spare buildings like the Louvre and the Elysée [the presidential palace], so that the bomb shall fall on *ordinary* people? It is painfully comic that Paris should have become suddenly so precious."[98] Others sought to make the best of the situation by organizing cellar parties that would serve as the occasion to share the best bottle in storage there.[99]

Fear did, however, grip some spectators of airship raids. Count Ottokar Czernin, the Austro-Hungarian ambassador to Romania, was interned in Bucharest at the time of three Zeppelin attacks in September 1916. He recalled the experience:

Last night a Zeppelin did come. About three o'clock we were roused by the shrill police whistles giving the alarm. . . . Suddenly darkness and silence reigned, and the whole town, like some great angry animal, sullen and morose, prepared for the enemy attack. . . . Fifteen, twenty minutes went by, when suddenly a shot was fired and, as though it were a signal, firing broke out in every direction. The anti-aircraft guns fired incessantly, and the police, too, did their best, firing in the air. But what were they firing at? There was absolutely nothing to be seen. . . . Was it really there, or was the whole thing due to excited Romanian nerves? . . . The searchlights sway backwards and forwards. Now one of them has caught the airship, which looks like a small golden cigar. . . . Smaller and smaller grows the Zeppelin, climbing rapidly higher and higher, until suddenly the miniature cigar disappears. . . . Suddenly utter silence reigns. Have they gone? Is the attack over? Has one been hit? Forced to land? The minutes go by. We are all now on the balcony—the women, too—watching the scene. Again comes the well-known sound—once heard never forgotten—as though the wind were getting up, then a dull thud and explosion.[100]

The initial confusion caused by the air alarm also complicated later investigations of the incident, as rumors spread and witnesses fed on one another's interpretation of the events.[101] In the morning, when the extent of the casualties was known, an outraged cry for revenge went up.[102]

In Paris, even apparently minor incidents could upset the population. Such incidents ranged from the single bomb that pierced the ceiling of a subway tunnel, killing no one, to the unexploded projectile that crushed a policeman as he prepared for bed.[103] Indeed, there were quite a few such incidents, for the bombs were remarkably inefficient, a fact that sometimes gave the Allies an opportunity to recover intact bombs, study them, and devise ways to defuse them.[104] Media reporting on these events nevertheless attempted to focus popular anger, exaggerating the morbid details and carefully listing the names of the deceased and the ages of the orphaned. Since the casualties were relatively light—not hundreds of corpses requiring disposal under war conditions—the few dead were given national honors, a fact that galvanized the population into holding up under subsequent heavy artillery bombardment.[105] It did not matter that the Zeppelins were actually quite inefficient; when a bomb killed a child, it was seen as evil proof of the precision for which the Germans were famed. Imaginations were easily stirred.

War often provides a stimulus to the creative mind,[106] and the airship certainly contributed its share to this phenomenon during the Great War. The sublime effect of the machine in peacetime continued to be felt in warfare. Artists and intellectuals were particularly taken with the contradictory mix of fascination and repulsion that the airship evoked. In London, while Arthur Conan Doyle raised the issue of retaliation, D. H. Lawrence mused on the parallels between the airship raids and the themes of Milton's *Paradise Lost,* and in *Zeppelin Nights* Ford Madox Ford focused on behavior under the blackout.[107] George Bernard Shaw, who opposed the conflict on pacifist grounds, complained to a German correspondent (who received his letter via neutral courier) that the only risks he faced came from Zeppelins dropping bombs without thinking that they might actually hit "the upright man."[108] He in fact remained amazed at the "impunity and audacity" with which the airships operated, but he also acknowledged that he found the machines so mesmerizing that he almost wished for another raid to occur.[109]

Others, however, saw airship raids as an odd sort of festival, in which all traditional social barriers were broken down and the population became divided instead between those who wanted to see the machines and those who rushed to the basement. To Marcel Proust, who described aeronautical anecdotes many a time in *Remembrance of Things Past,*[110] the Paris sirens warning of the arrival of a German airship could not have been a more appropriate, a "more Wagnerian," salute to the machine's impending attack: "It might have been the national anthem, with the Crown Prince and the Princesses in the imperial box, the *Wacht am Rhein,* one had to ask oneself whether they were indeed pilots [of intercepting aircraft] and not Walkyries who were sailing upwards."[111] Proust envisioned a technological ballet that entertained rather than terrorized, thanks to the apparent remoteness of the attack.[112] Jules Romain, in *Men of Goodwill,* preferred to emphasize the deadly efficiency of a Zeppelin almost killing a general in the rear.[113]

The reactions of intellectuals, although not always negative, combined with those of the general public to create a groundswell of potent antiairship feeling in France and England that would exacerbate the problems of the Zeppelin company after the war. Of course, the reverse was true on the other side of the trenches, where airship successes became part of daily war indoctrination.

Fantasy and Fact of Zeppelin Operations

In Germany, there was no better propaganda for the war effort than the sight of airships flying over the homeland on their way to a mission. The image suggested that the country was prepared for everything, as the media claimed, and it ensured a certain level of optimism. Some young men even joined the armed forces earlier than planned after experiencing the sublime effect of a combat airship flying over their region.[114] The number of airships downed, of course, presented a problem; losing an airship was different from losing a cannon or even a trench full of men. However, so long as a modicum of good news could be published, even these losses might be considered acceptable. Ironically, the cult of the airplane pilot,[115] viewed as competition in airship circles, may have helped offset the psychological effect of the losses, as the public discovered the likes of the Red Baron and shifted its attention to new sky-gods.

Hatred of the enemy also helped deflect frustration with the war. A rage against France and, especially, Britain had erupted in Germany following the outbreak of hostilities, but the emperor, afraid that his royal cousins might get hurt, initially refused to allow airship attacks on London.[116] However, calls for Count Zeppelin's "glorious" machines to bomb Albion began to be heard everywhere, even at the highest levels.[117] One commentary quoted H. G. Wells's *The War to End the War* (which imagined the destruction of Berlin), noting with relish that Wells's scenario was an "illusion" while the attack on Antwerp was a deadly reality.[118]

Street songs and popular rhymes celebrated the airship's destructive power. Newspapers were quick to laud the successes of this "most modern air weapon, a triumph of German inventiveness," and the German Airship Association's *Luftflotte* newsletter printed odes to Zeppelins alongside its monthly list of fallen airmen.[119] The Zeppelin raids on Paris were presented in the German press as "justified reprisals" for French air force bombings of German cities. Though the city was technically a civilian target, columnists explained, hitting it would demonstrate the triumph of German technology over French mediocrity. One such German raid reportedly included the dumping of leaflets stating, "Parisians! Here are your Easter Eggs!"[120] On the eastern front, airships had scored some successes in raids that destroyed railroad exchanges, but these results were as much due to the lack of proper air defenses behind the Russian lines as to the daringly low flying of the machines to ensure clear identification of the targets.[121]

German postcards often depicted Count Zeppelin or his machines. The German Airship League commissioned artistic portrayals of German successes, sold as postcards to support the war effort and reinforce nationalistic culture. During the early phase of the war, cards depicting the enemy population scurrying for cover at the mere hint of an airship visit were legion.

Books and illustrations intended for children, but obviously seen by adults too, reflected the spirit of defensive nationalism that had sprung from the fear of encirclement before the war.[122] In such stories, the use of the airship in support of the infantryman is justified by the threat of Britain's huge navy—depicted as a menace even though the war was fought mostly on land—and by Franco-Russian treachery over treaty issues. These representations celebrated both the airship and its inventor as uniquely German: "They do not have him and never will," stated a children's poem.[123] Count Zeppelin's exemplary courage would foil attempts at destroying Germany and fill its youth with pride.

The count, who had contributed to the myth of his machine as a form of ultimate weaponry before the war, continued to do so during the conflict. Although he admitted that the dirigible was slowly reaching obsolescence, in addresses to industrialists and politicians he advocated the use of airships in the conflict. "We stand in a fight for existence, thus we must use all means available to ensure our survival," he once argued, adding that the airship remained the most effective and least costly weapon toward the preservation of German identity. He had visited the German western front in late 1915, and what he saw there only strengthened his opinion that England should burn.[124] Conrad Haussman, the Reichstag leader of the Württemberg Progressives and an early supporter of a negotiated peace, recalled the count telling him in March 1916 that Germany should "strike England in the heart" by keeping the pressure up through air and sea attacks.[125]

This belligerence eventually backfired. The problem was the count's own popularity. He continued to attract interest wherever he traveled. His words were often quoted and his speeches reprinted like literary classics.[126] This fact worried the civilian government, which was trying to follow the "politics of the diagonal." Echoed time and again in various forums, the count's virulent views and his repeated challenges to the government to use the full potential of airships and airplanes were at odds with Chancellor Bethmann-Hollweg's attempts at negotiating various peace proposals.[127] The count had become a loose cannon, a menace to the very government that had held him up as the archetype of the good fighting German. By October he had publicly apologized numerous times for his behavior, hoping to mend his strained relations with the government.[128] There were, of course, no formal reprisals, but his situation vis-à-vis officialdom remained awkward.

It therefore comes as no surprise that Count Zeppelin's death on 8 March 1917 drew a mixed response. A state funeral took place, and all reports stressed that a genius had passed away. The count's eulogy emphasized his Christian up-bringing and military devotion to duty.[129] The front-line newspaper *Der Drahtverhau* (The Barbed Wire) printed a representation of Icarus taking flight in the shadow of an airship; the satirical *Simplicissimus* depicted St. Peter lead-ing the popular figure through the clouds and promising him a good view of Germany.[130] War conditions affected public memories of the count: for the most part, the hundreds of articles and obituaries published in March 1917 nos-talgically focused on the early days of his success, not his recent belligerence.[131] Not all the comments were favorable. The Baden writer Anton Fendrich drew a grim parallel with Fritz Haber, the father of chemical warfare: "Just like Count Zeppelin, [Haber] has mastered the air, only in a different way."[132]

Commemoration could not assume the same personality-cult dimensions as in peacetime. Oskar von Miller, who headed the board of directors of the famed Deutsches Museum, an institution dedicated to preserving Germany's scien-tific and technological achievements, sought to have a monument built to the count's memory, but the count's family and his business successors all objected on the grounds of the international situation. Alfred Colsman, one of his clos-est associates, felt that the public mood would not stand for such a tribute, and were any monument to go up, it should take the form of housing for widows and orphans.[133] In affecting the myth of the airship as a wonder weapon, the strains of war thus threatened the image of its inventor.

While Count Zeppelin's personal standing had suffered because of his con-duct before his death, propaganda literature continued to boost his machines' strategic bombings. Numerous war novels and supposedly factual accounts in-volving airships appeared, dedicated to Count Zeppelin or "the German people and German youth." These propagandistic potboilers and other works of pulp literature blended chauvinistic exaggerations with actual media releases to cre-ate an appearance of unshakable fact. Foreign press reactions were also often included in such works to convince readers that the fantastic claims of success heard at home were undisputed abroad as well.[134] Rumor had it that "anti-Zeppelin" airships were being constructed, and battles between them were de-picted in print as a kind of clash of titans.[135] It was an image that ignored re-ality: in fact the Allies were building airplanes that could climb higher and fire rockets at airships. Novels depicted the machines being used in a variety of roles, from the landing of troops to the rescuing of downed airmen. In this fic-tional world, airplane pilots helped their airship brethren—though in fact a traditional dislike divided the two groups—and airships conducting coastal patrols foiled attempts at dropping spies into Germany.[136]

The fascination with Zeppelins reached such a pitch that the public came to believe that any airship in the sky was a Zeppelin-built machine. This misconception proved most annoying to Johann Schütte, the engineering brain of the Schütte-Lanz airship company, which also manufactured bombers.[137] Media accounts often perpetuated the misunderstanding by reprinting the wires of the German WTB press agency, which routinely used *Zeppelin* to denote any airship: in these accounts, it was always "Zeppelin fire" that engulfed London.[138]

In most German accounts of the war, airships—in contrast to the raw mechanization of warfare on the ground—were giants capable of magic: enemy searchlights would seek them without success, their patrols would yield sudden ship kills, and the mere cry of "Zeppelin!" was sufficient to send everyone scurrying for cover. The reality of air combat was, of course, quite different. Postwar popular literature gave a skewed account of reality, and the memoirs of airshipmen did little to set the record straight.[139] In fact, the High Command had to learn to use the air weapon, as did those in the field.

Airship operations entailed a strange mix of technical protocols and artistic prowess, whereby each machine's success depended on the strength of its physical components as well as on the cohesiveness of the crew and the experience of the captain. In war accounts the crew remained nameless for the most part until after the war, although their work was often acknowledged as being tougher than that of seamen. Some airship captains—in particular Peter Strasser, the commander of the Naval Airship Division—were mentioned by name, and their relation to their machines was likened to that of a horseman to his mount (airplane pilots were described in much the same way).[140]

Such proud stories masked the difficult reality of airship flying. Airship crews were trained to trust their machine, but each group's behavior varied according to its captain's leadership style and work practices. In the tension preceding a mission, one commander would obsessively order every ballonet checked. Another, buoyed by the luck he had experienced thus far, might dispense with the full series of checks: "We had such confidence in it," noted one such commander, "that I declared LZ 79 ready for departure."[141] In either case, given the complexities of ground handling, it took up to two hundred men to steer the airship on the ground before takeoff and after landing.[142] The parallels with seagoing ships were unmistakable, and although there were no prescribed rituals on board, informal ones existed, such as the flying of a regulation military flag from the tail during early operations.[143]

While takeoffs occurred in daylight, raids usually took place at night, during the dark of the moon. Airship tactics were thus complicated not only by the High Command's inexperience with the air weapon (which included ignorance about what to bomb) but also by weather conditions and the problems of night aerial navigation. To allow airships to benefit from cloud cover yet be able to pinpoint their location, several machines received a tethered, streamlined observation car, in which a crewman could be lowered below the clouds. The steel tether, approximately 750 meters long, had a phone line attached to it so that the observer could call for the bomb drop. The contraption was first used in March 1916, when Calais was bombed, but throughout the war it gave only mixed results. Despite the cold, crewmen volunteered for duty in the car, as it was the only place they were allowed to smoke.[144]

Aboard the airship, crew worried constantly about engine breakdowns, hours of cold discomfort, and the possibility of being burned alive should Allied planes shoot at them.[145] Lack of oxygen was sometimes a problem as well. When the bombs were released, the airship would suddenly gain altitude and, depending on the height, incapacitate the crew by exposing them to the rarefied atmosphere. Despite recommendations that crew members inhale bottled oxygen in such situations, many refused to do so for reasons of comfort (prolonged use cracked the skin and caused a hungover feeling) or out of fear that their manliness would be questioned. It was only once they noticed problems with crew performance under oxygen deprivation that airship captains moved to enforce the regulations. Captains also faced new challenges in the form of weather predictions that held for low altitudes but became useless at higher ones. Finally, navigation above the clouds remained a nightmare.

No matter how experienced the crew was, a single patch of fog could disrupt operations. True, airshipmen might thus gain the advantage—in fog, the enemy could not see them, either—and use radio beacons along the coast to determine their position. Yet bad weather could also interfere with radio contact, and there was the risk that the enemy would intercept the beacons. The return to land might become as nerve-racking as the bombing raid had been. Often, at the base, the call "Something from above is here!" (signaling an airship arrival) would go out to ground crew, yet nothing would happen for several hours. Around daybreak, as the surrounding relief became clearer, airships would attempt a landing. The combination of crew fatigue and poor visibility sometimes caused a rough landing or an outright crash.[146] Despite such troubles, the belief that these problems would soon disappear remained strong at the rear, thanks in part to propaganda.

The illusion of success that characterized the first two years of war is best exemplified in the nine-airship raid launched against England in January 1916. The event received considerable press coverage despite the overall failure of the mission and the loss of one machine (the news was kept from Count Zeppelin for several days).[147] The results of the raid indicated that the airship commanders had been unable to navigate their machines; the details were reminiscent of LZ 16's unintentional arrival in Lunéville before the war. Hugo Eckener, a close associate of Count Zeppelin who was serving as an airship instructor, blamed a military structure that offered command to inexperienced officers, overlooking obvious choices among older ranks. (Eckener would invoke this lesson after the war to require that civilian captains have at least four years of airship experience.)[148]

Even the commanders who missed their targets still received honors upon their return. As for those who did not return from a mission, rituals were devised to honor their memory. Early in the war, postcards representing the heroic dead were printed, as if to offer the public new martyrs to worship. Closer to their home base, slow waltzes played on gramophones to commemorate the dead. Eventually the records spun so often, Eckener gloomily wrote his wife, that it was not worth getting to know any of the new airshipmen, "for their lives were so short."[149]

In contrast to the general public's ready acceptance of the airship as a wonder weapon, many officials were apparently aware of the dirigible's inefficiency and shortcomings as early as 1915. As airships began to suffer defeat, some German observers became concerned that public opinion would turn against the government.[150] The impact of the bombings on public opinion in neutral nations also worried Chancellor Bethmann-Hollweg. Nevertheless, the official line was that the machines were valuable reconnaissance tools, and that together with the submarine they would ensure victory over England. A skeptical Count Hugo von Lerchenfeld, the Bavarian state representative in Berlin, remarked to his superior in Munich that this claim was misleading; in fact, officials accepted the Naval Airship Division's arguments only because there was definitely no other way to take on and beat Great Britain.[151]

More tepid arguments from military leaders claimed that airship bombings had forced Britain to maintain on the home front "a large number of troops, artillery, and ammunition, which would otherwise be fighting us at the front."[152] This assertion, which would become the stuff of myth in postwar memoirs, is contradicted by contemporary British accounts; and indeed, the point was moot, since trench warfare had already changed the face of war and the airplane was beginning to gain new ground. General Paul von Hindenburg later claimed in his memoirs that during a visit from Count Zeppelin in 1916, the count had acknowledged the obsolescence of his machines and was ready to focus on airplane design exclusively.[153]

German propaganda continued to ignore the airship's shortcomings and presented exaggerated claims of destruction in the reports it dispatched to the press.[154] Naval airship commander Peter Strasser and others like him deliberately ignored the setbacks and worked obsessively to perpetuate myths about airshipmen paralleling those of knightlike airmen pushing their machines to the limit.[155] But in fact, there appeared to be a lack of direction at the higher levels of command concerning airship matters. Whereas early in the war, Berlin bureaucrats had issued specific orders, by 1917 the Admiralty tended to rubber-stamp Strasser's reports and demands, to the point where he found himself obsessively devising plans of his own until his death.[156]

Strasser's fiery demise in one of the last dirigible raids on London on 5 August 1918 also reinforced the airship myth. After four years of war, the loss of L 70 barely attracted attention, even though its effect was to stop all further bombing missions (sea patrol missions, however, did continue). Yet the myth lived on. By the tenth anniversary of his death Strasser was remembered as a genius whose successful efforts against England had also, by perfecting the technology, contributed to making Count Zeppelin's creation a wonderful instrument of peace.[157]

On 9 November 1918 the German Empire fell, and two days later the Armistice was signed. In the early days of the Weimar Republic, the revolutionary mood rejected the nation's imperial heritage, which included the airship. Military leaflets that attacked Admiral Alfred von Tirpitz's leadership of the German fleet, for example, also openly questioned the value of the dirigible in terms of both technical and military achievement; of the 117 German airships operated during World War I (Zeppelins as well as other companies' machines), 39 were shot down and another 42 were lost to other factors, from weather to technical flaws and crew error. Yet the rebuttal was not long in coming. Although the matter was something of a tempest in a teapot, representatives of the Zeppelin company, in response to an outcry in the press, published a "counter brochure" defending the airship.[158] Their defense emphasized the nationalistic and technological heritage of airships. Yet the need to defend the company's good name in the first months of peace paled in comparison with the challenges of the clauses of the Versailles treaty, which threatened its very life.

The Versailles treaty added to the other challenges already affecting the Zeppelin company. First, Zeppelin's production of airships and airplanes ceased at the end of the war, when the revolution interrupted fulfillment of military orders.[159] True, some of the Zeppelin factories could be sold or converted, but there remained the question of employing the people working there. There was also disagreement within the company as to whether to convert to the manufacturing of other industrial products: Alfred Colsman, who had helped Count Zeppelin since the early experiments, advocated diversifying into the household-appliance field, while Hugo Eckener, also on the board of directors, clamored for staying true to the count's spirit and continuing to concentrate on airships.[160] In hindsight, it is clear that the Zeppelin company's survival during the interwar period owed much to Eckener's public-relations flair and Colsman's wise investments; but for the moment the firm faced more pressing problems than the postwar economic downturn.

Under the provisions of the Versailles treaty, the Allies required all German aeronautical production to cease and all airships to be transferred, even those the German government insisted would be used exclusively for civil air transport. Several civilian projects were already being planned in Germany, new discussions about dirigible technology having arisen even as the failure of the wonder weapon was becoming clear. As early as 1916, talks concerning a possible airship link between cities had taken place.[161] The African expedition offered additional stimulus to such ideas, leading some to hope that Germany, which had been ahead of all nations in airship design during the war, would remain so in peacetime.[162] However, the Allied conditions remained clear and inflexible: the dirigible's failure as a strategic bomber did not hinder its potential as a military reconnaissance machine, and this troubled the Allies.

Several attempts were made to mollify Allied attitudes. A group of German airshipmen working with Zeppelin representatives planned a nonstop Atlantic crossing to sell one of the war machines in the United States; technical problems and trouble with the negotiations canceled the project.[163] Another group made a sales pitch during the surrender of a machine to England and suggested that commercial airship transport would ensure the Zeppelin company's survival; those claims fell on deaf ears.[164]

In September 1920, in response to Allied demands, all German military airship units were officially disbanded after thirty-six years of existence. Following the example of the scuttling of the German fleet at Scapa Flow, angry airship crews destroyed seven airships scheduled for delivery to the Allies.[165] This action prejudiced the Inter-Allied Commission against the Zeppelin company, thus complicating any plans for future transport airships and shaping the activities of Hugo Eckener during the Weimar era; many of his early efforts would revolve around rebuilding a modicum of trust for the company's civil and commercial designs.

Upon receiving news of an airship raid on London, Thomas Edison had argued that the next war should be fought with machines instead of men.[166] He suggested that such an approach to war would offer efficiency and also have a deterrent effect. Dirigibles did neither. The airship played a complementary role in the first total war as one of many new technologies to appear on the battlefront. Initially a "peripheral weapon" that received backing much later than its supporters had hoped,[167] it nonetheless underwent substantial technical improvement during the war, partly in response to new conditions. New engines, more efficient lift ratios, navigation tools and radio[168]—all contributed to the impression that eventually the Zeppelin's purported superiority would be proven. The technology came very close to maturity, as will be seen later, but none of these improvements broke the stalemate of the trenches; it took economic exhaustion, combined with another technological god, the tank, to do that.[169]

Despite its fall from military grace, the airship remained an intrinsic part of German culture and the German war effort. The favorable public and media perceptions of the machine at war also affected perceptions in the government, which maintained the illusion of airship success for a long time. Thus, the Zeppelin sublime effected the creation of a new myth: that of a technological skygod in a secularized "holy war" between Germany and its enemies.[170] Count Zeppelin had from the very beginning of his experiments participated actively in the propagation of this image. Even between 1908 and 1914, when his machines assumed other functions as commercial transports and as malleable symbols of peaceful patriotism, he did not waver from his initial goal of using them to give his country an element of military superiority. Theodor Heuss (later the first president of the Federal Republic), who had written enthusiastic columns on Count Zeppelin's 1908 endeavors, reflected bitterly on the count's obsession: "He disappointed us."[171] Yet after the war, and despite an initial rejection of the Count Zeppelin heritage, nationalist accounts of commercial airship activities would emphasize the debt owed to his heroic efforts and to the crewmen who had served in the conflict.[172]

In England and France, on the other hand, popular arts, propaganda, and memory produced the reverse effect. The estimated four thousand casualties airships caused on all fronts were but drops in the ocean compared with the dead from Verdun, Caporetto, and the like. Yet the hatred the military airships evoked as the first strategic bombers continued to strengthen the myths that had kept them active throughout the war. Fact and fiction about them mingled freely in histories of the conflict, and few civilian memoirs of the war fail to mention the rumbling monsters whose sublime effect both fascinated and terrorized. Such symbolism may not be discounted in considering the Allies' decision to destroy the entire German airship industry. For Germany to save it would require political acumen, exploitation of the machine's emblematic heritage, and careful manipulation of nationalist feelings.

CHAPTER 4

The Airship as a Business Tool in Weimar Culture

I N THE FIVE YEARS THAT FOLLOWED THE END OF WORLD War I, Germany plunged into social, political, and economic chaos, experiencing only a brief respite of stability between 1924 and 1929. With the stabilization of the Weimar Republic, however, Germans rediscovered their fascination with the Zeppelin phenomenon. Many national icons had disappeared when imperial Germany fell apart, but the resurrection of the airship offered Germans the double opportunity to reminisce about a bygone golden age while looking toward a modern and better future. The Zeppelin's rediscovery was tied directly to the expectation that it could contribute to improved means of transportation, to better commercial links, and to the restoration of Germany's honorable reputation. Eagerness for this kind of progress was a facet not only of the maturing passion for modernity but also of the wish to see the nation continue contributing to aeronautical history in the face of the restrictions imposed by the Versailles treaty.

In 1922 the artist Laszlo Moholy-Nagy remarked that "the reality of our century is technology. . . . To be a user of machines is to be of the spirit of this century."[1] The very embodiment of modern technology, the airship was an apt symbol of the so-called *Neue Sachlichkeit,* or New Realism, that permeated the art world of the 1920s. This movement represented a coming to terms with a new phase of modernity: sobriety was to replace the ecstatic, function was to offer a new objectivity. The corollary to the embrace of the machine, as Moholy-Nagy saw it, was that technology displaced tradition. In the case of the airship, however, the sense of national tradition was actually amplified by business interest in the technology; as a commercial embodiment of the technological sublime, the Zeppelin became a symbol that merged tradition with progress.

The airship exemplified a complex new mass culture that came to the fore in the 1920s, further blurring the line that separated Germany's classical tradition from its modern heritage. Popular German literature of the era abounded with icons of progress.[2] Imagery mattered, and the occasional appearances of airships during Weimar Germany's existence fit the bill for an avid media and a public eager to consume such images. But however clearly the airship seemed to represent progress, its success also depended on public awareness of its imperial heritage. The reception of the airship echoed the tensions that characterized Weimar attitudes toward technology. Despite the horrors machinery had caused during World War I, the New Realism welcomed perfected technology. Perfected though they were, Zeppelins were reaching twenty-five years of existence and entering a new political era. That their reappearance in German skies caused an explosion of generational nostalgia and youthful exuberance is due in part to nationalist circumstances, but also to the remarkable acumen of one of the Zeppelin company's directors, Hugo Eckener.

Rescuing the Airship from Oblivion

Hugo Eckener, the main figure in the Zeppelin revival, trained in psychology at the turn of the century, then made his living as a journalist. Eckener first met with Count Zeppelin after having written a scathing article about the airship. Eventually the two became friends, and Eckener went to work for the count, also learning to fly airships. During World War I, Eckener trained airship crews, earning the nickname "The Pope" for his strong resolve and doctrinaire approach to teaching. At the end of the war he returned to Friedrichshafen. A political conservative, he nonetheless espoused certain liberal ideas. He would thus rely heavily on mass culture to achieve economic success for the airship. Several years passed, however, before he could do so.

With the cessation of hostilities in November 1918, the voiding of military contracts drove the Zeppelin company to the brink of bankruptcy. Even deliveries in progress came to a halt for lack of payment. In addition, Zeppelin manager Alfred Colsman had to navigate the rough seas of revolutionary upheaval, as factory workers struck on and off through 1919.[3] In the meantime, the Zeppelin concern required new means of survival. One option common at the time was to increase diversification while letting go of various subsidiary companies so as to increase liquidity. The Zeppelin firm moved to manufacture gearboxes, automobiles, and even aluminum pots and pans. Supporters of the airship idea, however, felt that airships alone, rather than a mix of diverse products, should sustain the company.[4]

One obvious way to ensure the survival of airships was to reestablish civilian air links, at least along the lines of those that had existed before World War I, and then extend them through plans drawn up during the war. In early 1919 Colsman and the board of directors accepted recommendations to begin running an airship line between Friedrichshafen and Berlin. The Versailles treaty was still under review, which meant that air transport remained a viable option. The company resolved the matter of hardware by building a new airship that took into account technical progress made during the war.

The new machine, LZ 120 (later christened *Bodensee*), represented a new stage in airship building. The great airships *Los Angeles, Graf Zeppelin,* and *Hindenburg* would follow its basic design. Credit for LZ 120 goes primarily to Zeppelin engineer Paul Jaray, whose brilliant calculations of optimal aerodynamics went mostly unnoticed during the war. Once the project team had received the go-ahead, Jaray produced outline drawings in a matter of weeks, and construction was completed by August 1919.[5]

To the uninitiated, the machine appeared ridiculously small and plump. It was indeed the shortest of all the Zeppelin airships built, even after a redesign extended the fuselage slightly. The lack of a tubelike section in the middle helps explain the uncommon appearance. The conservative Zeppelin approach to airship design had long assumed that such a section was necessary to help the airship maintain effective steering. Jaray both disproved this and demonstrated that a teardrop shape would increase speed. As for the cabin, it represented a third-generation design. The first arrangement, used until 1914, consisted of

The LZ 120 *Bodensee* lands in Friedrichshafen in the fall of 1919. The airship's profile shows clearly the new teardrop design engineer Paul Jaray introduced with this machine. Successful on the Berlin-Friedrichshafen link, the *Bodensee* was withdrawn from service in compliance with the provisions of the Versailles treaty and handed over to Italy. LZA, 120-0105

two open suspended nacelles, sometimes with an enclosed passenger section in between. The second, applied in wartime, featured enclosed compartments, still suspended. Now, in LZ 120, the cabin was attached directly to the fuselage, thus improving streamlining. The *Bodensee*'s top speed, up to 130 kph, was double that of prewar transports. Even when carrying a full payload—between fifteen and twenty passengers, depending on distance and weather conditions—the airship covered the distance between Berlin and Friedrichshafen in nine hours at most; a strong tail wind once cut that time by half.

The Zeppelin company used both speed and the very experience of flying as selling points. Its advertising brochures contained descriptions of the beauty of landscapes similar to those of the prewar DELAG's pitches to potential passengers. Comfort was comparable to that of trains, for the designers had taken special care to install light wicker chairs covered with red upholstery. A menu of cold foods was offered—warm drinks were served from thermos bottles—and no smoking was allowed. For the first time, and in stark contrast to airplanes, passengers could use specially designed toilets. The lack of a heating system, though, made fall flights extremely uncomfortable (the toilets froze, too).[6] Despite these and other shortcomings, passengers responded positively to the service: although expensive (475 marks, or $15), tickets required several weeks' advanced booking.[7] Yet there arose no Zeppelin craze like the one that had existed before the war. The harsh economic situation and the humiliation of Versailles were just beginning to sink in, and Germans had other concerns besides applauding a machine that reminded them of imperial times.

The Zeppelin company had planned extensions to Stockholm too, since only international flights to former war neutrals would be possible. A second airship, the LZ 121 *Nordstern,* was completed in the winter of 1920 for that purpose. Unfortunately, Allied pressure to enforce the clauses of the Versailles treaty canceled the link.[8] It was suggested that Switzerland might intervene, together with Sweden, in Germany's behalf, but nothing came of this idea.[9] The machines left Germany for service in France and Italy as compensation for the seven airships of those nations wrecked at the end of the war.[10]

The requirements of the Versailles treaty called for the destruction of all German aviation installations, including the Zeppelin factory in Friedrichshafen. Not surprisingly, these requirements met with widespread public opposition in Germany and in some circles abroad.[11] However, such outbursts made little impression on the Allies and offered no practical solution for the future.

One alternative, to seek out financial support from another nation to start building airships anew, focused first on France (a fact that observers naturally ironized about), then on Spain. As a departure point for links over the South Atlantic, Spain offered ideal conditions for continuous winter operations, whereas in Continental Europe, all regular air links were closed between December and March. The intricacy of Spanish politics and the need for heavy investment slowed the company's efforts, but the project continued to occupy Colsman and Eckener for years.[12]

Then another nation—the United States—stepped into the fray to assist in the Zeppelin company's recovery. The history of American interest in German airships is an intricate one, with the United States having been in the paradoxical position of encouraging Germany to further develop Zeppelin technology while at the same time working with the Allies to enforce the restrictions the Versailles treaty placed on airship construction.[13] In 1922, after a round of failed negotiations with the U.S. Army as well as a series of dealings with the industrialist Henry Ford in 1920,[14] Zeppelin representatives secured a tentative contract from the U.S. Navy for the construction of a new and bigger transport airship.

The road to success was not easy. Germany had to negotiate with two Allied bodies, the Inter-Allied Air Control Commission and the Conference of Ambassadors,[15] an organ established in Paris to monitor German implementation of the treaty and payment of reparations. Each side had its own priorities. To the Germans, the value of building one new airship was that it would keep the Zeppelin company alive and thus preserve Germany's "industrial potential." The company, fearing that trade secrets might be taken away, would not agree to Allied control of the construction, let alone delivery of plans for construction elsewhere.[16] Eventually an agreement was reached with the behind-the-scenes support of the German foreign minister, Walter Rathenau. The AEG industrial concern, which Rathenau had headed during the war, certified the "financial health" of the Zeppelin company, whereupon the contract was prepared.[17] In taking this step, Rathenau put national business interests ahead of personal considerations: he ignored the unfortunate incident that had involved his father and Count Zeppelin in August 1908, for the sake of preserving the industrial potential of the company. Unfortunately, Rathenau did not live to see the agreement implemented, as he was murdered by right-wing fanatics two days prior to its signing.[18]

The agreement also inaugurated a new era of subsidies on the part of the German government for Zeppelin. Some 3 million gold marks were advanced to the company for the construction of the airship,[19] thus signaling the beginning of an industrial policy of supporting the remnants of the German aeronautical industry. The only shadow looming was that of a patent fight with airship competitor Schütte-Lanz.[20] A thorn in the Zeppelin company's side, the dispute was eventually resolved some years later, but it points to the multilayered domestic and international problems the company faced in its attempts to bring the airship back to life.

With the new airship contract, the Zeppelin company found itself on a double track: airship building and diversification. Alfred Colsman, focusing on the bottom line, continued to advocate the second path. He argued that even though the initial purpose of the company had been "purely patriotic," the need to preserve Count Zeppelin's legacy in one way or another meant that he had to consider such compromises.[21] Others, however, wanted to bring the airship back to the fore. In an internal memorandum that quite overlooked Count Zeppelin's own martial opinions, his nephew, Freiherr von Gemmingen, claimed that the count had always meant to use his invention as a cultural tool to establish transoceanic links but that the war had disrupted these plans.[22] It was now up to the count's successors to bring about a rebirth of the airship in the interest of *Kultur,* and thus rehabilitate both the count's memory and Germany's reputation. Colsman conceded that German public opinion might misconstrue the company's moves toward diversification and internationalization as evidence that its leaders were motivated by purely economic interests rather than patriotic ones. The Foreign Ministry was also dubious about internationalization: to serve German business interests and to honor the "German spirit of enterprise and inventiveness," an international move should be undertaken only as a last resort.[23] In the meantime, the U.S. Navy contract would have to do.

Construction began in 1923. A joint venture with the Goodyear Rubber and Tire Company followed a year later. These events coincided with the beginning stabilization of Weimar Germany's political and economic systems. Precarious though these systems remained, they were stable enough to permit the German public to focus on something other than social welfare and savings, like a new airship.[24] After all, the idea of flying a dirigible across the ocean was front-page news. The British airship R 34 had completed a successful transatlantic round trip back in 1919. That airship had been a loose copy of a German machine, and now a few German columnists quipped that R 34 had convinced the Americans of the quality of German engineering.[25] At the time, the flight had attracted only limited attention in Germany, partly because of the purely military nature of the mission, and partly because of other events dominating the news. But now, five years later, the planned Zeppelin transatlantic flight was a matter of both technical interest and public concern.

To the German public, building a Zeppelin was a good thing; that it would fly across the Atlantic, even better. However, the question of national property was on everyone's mind. "What is it with this airship built in Germany by German workers and engineers, paid for with German money, but which belongs to America?" clamored a Berlin columnist.[26] It was bad enough that the aeronautical industry had almost disappeared five years earlier, but now a favorite cultural icon was to be placed at the service of a victor nation. The international political implications aroused suspicions. Conspiracy theories abounded, one going so far as to claim that the U.S. Navy's designation for the machine, ZR 3, meaning "third Zeppelin, rigid," stood in fact for "third Zeppelin, reparations."[27] Distrust ran so high that when the company postponed test flights and the delivery date, some felt that even this was the result of an Allied conspiracy. Others suspected that the company was simply stalling for time, since the project was only a stopgap measure to keep the company from

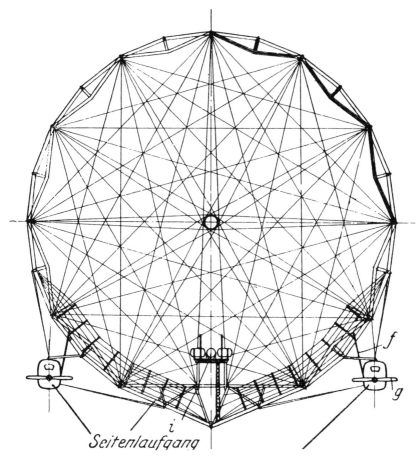

The improved center cross-section of LZ 126 or ZR 3. The reparations ship's design included several improvements based on technical lessons learned during World War I and applied in part to the design of the transport airships LZ 120 and LZ 121. Here, engineers Ludwig Dürr and Karl Arnstein ordered the addition of internal aluminum supports (see darkened sections). Each girder on the outer ring had a mirror opposite, which contributed substantially to improving the airship's solidity. The lines drawn across the fuselage section represent steel wires linking all sides of the tube and separating sections where the gasbags were housed. Ludwig Dürr, *Fünfundzwanzig Jahre Zeppelin-Luftschiffbau* (1924)

closing. Colsman had some difficulty convincing the Foreign Ministry that the work was encountering genuine technical difficulties.[28]

By August 1924 the machine at last stood ready for its first flight. LZ 126, as it was also designated by the company, was designed to fly as a military transport airship. Apparently just a bigger version of the *Bodensee*, it sported several improvements. Electrical systems included lights, a powerful radio, heating, and even a stove in the kitchen.[29] The hull's outer rings presented a novel arrangement. Until now, the hull rings had been assembled as polygonal structures, relying on their geometry and on the tensile strength of duralumin for resistance to inner and outer forces. This time, every girder was fitted with a "backup" facing inward, like a mirror image, so that the cross-section looked like a series of diamond-shaped structures.[30] Not only did this solution increase the hull's outer resistance, but it ensured that the gasbags netted inside the rings would not stress them. The hull itself received a coat of aluminum-colored dope to protect the interior from temperature variations caused by the sun.

The five engines were giant twelve-cylinder types, 950 kilos heavy. The biggest ones until now had been six-cylinder Maybachs—weighing far less but also consuming far more, for a diminished return. The new engines were reversible, allowing for quick stops but also saving weight on the bulky reversing gear that had been necessary until now. This concept, an American development from World War I, had never before been applied to airship engines.[31] Applying it had been the real source of the delays in the airship's construction. The Zeppelin company blamed these delays on both the new nature of the engines and the need to run them for hours without relief, so as to ensure smooth operation over long distances. But now the technical problems had been ironed out, and with the installation of aerodynamically shaped Jaray propellers, everything was set for long-distance testing, including a flight over Germany.[32]

The new airship's endurance flight over Germany had a cathartic impact on the nation. Feelings of guilt and wounded pride accentuated by economic troubles yielded to a sudden surge of public joy and national self-esteem. The media avidly interviewed Germans who remembered Count Zeppelin's 1908 success in overcoming both political and popular apathy. Attempting to clear away any remaining anti-Prussian feelings, the Berlin establishment's *Vossische Zeitung* reminded its readers that northern Germans had as much respect for the count's work and legacy as did his southern German compatriots. The Social Democratic *Vorwärts* echoed the theme of national unity under the airship banner.[33]

Now called by its American designation of ZR 3, the new airship demonstrated once again that while German technology may have faced difficult beginnings, its enterprising spirit always led to success. ZR 3 was no exception; it too would prove the quality of German workmanship. The good wishes accompanying the airship came from all across the political spectrum.[34] *Vorwärts* went so far as to equate French opposition to the Zeppelin company with the capitalist designs of the Krupp concern, the Social Democrats' archenemy. One needed to understand that crossing the ocean was purely a "cultural feat" (*Kulturtat*), to be accomplished with the fruits of cultural progress.[35] The only notable opposition to the achievement came from the German Communist party, which viewed the machine as a resurrection of German imperialism. Strangely, Communists and ultraconservatives agreed on one thing: the Zeppelin company was no better than a traitor to be handing over the product of "German workers' sweat" to the Americans.[36]

One way or another, then, the airship had once again become a uniting symbol for Germans of every political orientation.[37] The correspondent for *Le Journal* described the mood of Berliners as ZR 3 cruised over their city. Performing as if it were the main actor in a play, the machine had turned roofs into auditoriums. Spectators broke into tears of joy at the sight of it, while one teacher had his pupils salute it with a singing of the national anthem. The French journalist acknowledged that "this deep emotion was justified. . . . After five years of political chaos and economic misery, Germany has found itself again."[38] Reminiscent of the earlier reactions to airship overflights, ZR 3 brought back a sense of security with a vision of a golden past when political and economic stability had seemed assured. The vision was more than escapism; the Zeppelin's flights proved a capacity for success that many had feared lost. *Simplicissimus* captured the feeling in a drawing of two children bidding farewell to the airship with a song inspired by a seventeenth-century original:

> Fly, Zeppelin, Fly
> Father was in the War
> Mother was in the Homeland
> The Homeland burned down
> Fly, Zeppelin, Fly![39]

Despite the constant awareness of recent chaos, descriptions of the machine as a messenger of peace became a central theme in many interpretations of the event.[40] President Friedrich Ebert sent a congratulatory telegram that called for "free and peaceful competition among all peoples."[41] The airship was no longer a tool of war, although its creator remained a symbol of strength and endurance, as witnessed in various soldiers associations' bulletins.[42]

Notable dissenting views did, of course, exist. The ultranationalist *Deutsche Zeitung* saw the Zeppelin overflight of France on its way to America as "repayment, but not yet revenge."[43] Karl Arnstein, a Zeppelin engineer who had helped design some sixty airships, became the target of antisemitic comments: in view of his contributions, he might be only a "half-Jew after all," but the tarnish left by his work on the German symbol was indelible. Moderate conservative and socialist papers alike attacked such commentaries.[44] The overall effect of ZR 3's flights over Germany and across the Atlantic was to create a glimmer of hope embodied in a tangible form—in a machine that all could see, rather than in the behind-the-scenes efforts of German officials to maintain stability in the nation's domestic and foreign policies.

With the construction and delivery of ZR 3 to America in October 1924, the German government once again became involved in airship matters. Just as before the war, politics could not ignore cultural symbolism, especially where technology was concerned. From 1924 on, several ministries would regularly occupy themselves with the activities of the Zeppelin company. These dealings did not always go smoothly, and the Foreign Ministry frequently had to intervene to smooth ruffled feathers abroad.

As an example of the Foreign Ministry's concern with German airship matters, one need only consider the attention attracted by an article on the legal implications of the transatlantic flight of ZR 3.[45] The Foreign Ministry reacted warily to the prepublication draft, whose author, a high-level official in the Transportation Ministry, called for the right to build airships of any size.[46] It was feared that the piece might antagonize the Ambassadors' Conference in Paris, which had consistently forbidden any exception to the rule that Germany could build only small airships.[47] The Foreign Ministry was concerned with this apparently minor issue—once published, the article in fact caused no big splash—because the German government's interest in the airship issue went beyond helping the Zeppelin company; it involved wider concerns of overturning clauses of the Versailles treaty. To the Allies, ZR 3 was but a temporary reprieve, after which the injunction to destroy the airship works and hangars would resume. But as the German transportation minister later pointed out, the destruction of the Zeppelin operation would be contrary to the interests of all concerned. Spelling disaster for the nascent air transport infrastructure, it would hinder the recovery of German business, "in the sense defined by the Dawes plan," and therefore Germany's ability to pay war reparations.[48] Foreign Minister Gustav Stresemann's public statements echoed this outlook, and he brilliantly manipulated the issue of foreign ownership of ZR 3 to that effect, linking it with wider concerns of civil aviation and culture:

When we are required to authorize the overflight of Germany by a foreign airship, which in the interest of the development of this new means of transportation is desirable, then one must also promote the development of German airplanes, which is wrongly held up. What Germany could contribute to this field is demonstrated by the exploit of the new airship, which Germany eyes with pride. We are often asked in complaints why we attack the spirit of Versailles so passionately. Germany is building the biggest and most successful Zeppelin ship, which shows human development a new way, so that it may then be forced to tear down the airship hangars in which this result of the triumph of human spirit and human technology was born. That is the spirit of Versailles that we oppose.[49]

The message became the leitmotif of official and unofficial German views on the Zeppelin: no longer an instrument of war, the airship would now contribute to the overall mission of humanity by establishing air links that would strengthen world culture.[50] Conservative members of the Reichstag maintained that "the age of world air transport . . . cannot do without the inventive strength of German engineers, nor the quality work of German workers."[51] The Zeppelin was an essential instrument in the "international progress of the transportation arts, with their implications for civilization and peace."[52]

By publicizing the matter, the German government deliberately challenged the Allies. Should the former enemies decide to enforce their edict, the matter of prestige could easily be raised to the point of creating "a popular movement for the conservation of the airship works, so that not one German hand would be available to help with the tearing down."[53] The Foreign Office, for one, trusted that matters would not come to this. It had high hopes that by linking airship hangars to regular airship operations, and by working on the Allied nations individually, it could save the dirigible industry. Support from the United States, where sympathetic comments appeared in the Hearst newspapers, and the swaying of such British personalities as the air minister, Lord Thompson, led to the belief that this goal might be reached, especially while world public opinion remained enthusiastic about the successful ZR 3 transatlantic crossing.[54]

The successful completion of the transatlantic flight signaled a new era for Germany's international reputation. Whereas Albert Einstein's newly proven theory of relativity was often praised but poorly understood, the success of the airship provided obvious and tangible proof of the value of German science and technology. Congratulatory telegrams poured into German embassies.[55]

A Zeppelin fraternity. Some of ZR 3's crew pose around their captain, Hugo Eckener, following the transatlantic delivery flight of the reparations ship. Several of these crew members would later command commercial airships. *Left to right:* Albert Sammt, Leo Freund, Max Pruss, Hans von Schiller, Hugo Eckener, Anton Wittemann, Hans Flemming, Walter Scherz, Hans Ladewig, Willy Speck, and Ludwig Marx. UTD, Rosendahl papers, box 17, folder 14

Knowledgeable aeronautical journalists, recognizing the flight as a business venture, judged it to be the "opening of transoceanic air traffic" and belittled as a mere "sports feat" the success of the British R 34 airship five years earlier.[56]

The economic argument remained at the center of the international airship debate because the press echoed the theme so regularly. Nevertheless, the effort remained an uphill battle for Germany. To the Allies, and especially to the French, this constant emphasis on the international economic advantages of the airship appeared to be a diversionary tactic to push back the peace treaty conditions.[57] Some of Germany's neighbors, however, expressed more positive opinions. In Switzerland, for example, public opinion deplored the "senseless treaty" that impeded a barely tested technology that deserved a chance in view of the advantages it offered humanity.[58] Austrian comments were openly enthusiastic and emphasized the pride of being part of the German people, and thus of the German technological spirit. The flight of ZR 3 was "a symbol that Germany would reach its standing in the world through peaceful work," claimed the Austrian vice chancellor; to destroy the Zeppelin factory would be "a cultural ignominy."[59]

Similar words were on the lips of many German expatriates, too—the ethnic Germans in every part of the globe who were subjected to the wave of anti-German sentiment that surfaced during and after the Great War. The Reich government was concerned about Germans living abroad, and so were many private and religious organizations. Together such groups worked to provide an increased sense of security to expatriates, to give them a voice, and to show them that German culture remained strong despite the lost war. The support offered was sometimes quite concrete—for example, groups abroad were secretly financed through the German Foreign Office[60]—but the German colors always had to be displayed in a manner that would not compromise the fragile balance with the Allies that Gustav Stresemann had achieved by 1925. Among the private organizations that flowered during the Weimar period, several sought to glorify the airship as a traveling cultural heritage; German communities abroad seemed to lack a sense of cultural pride, and the airship was an ideal antidote. The pride campaign succeeded almost too well, because the goal of overturning Versailles was never far from view. For example, the German "Save Our Honor" League, claiming the support of "thousands of nonpolitical organizations," contacted the Ecuadoran government in the hope of persuading it to raise its voice in favor of German dirigibles in particular and aeronautical construction as a whole.[61] Such naive expectations indicate the lack of political control available to Weimar authorities. German associations abroad, in particular in South America, proved quite helpful for German business, but they were often a headache for government offices.[62]

Not all the developments in expatriate communities posed problems at home. Spontaneous celebrations of the airship by Germans around the world offered a good opportunity for Germany to patch up relations with the host nation, thus confirming the value of the dirigible as a goodwill ambassador.[63] On the one hand, the Foreign Ministry kept forwarding requests for airship overflight clearances to the Transportation Ministry and to the Zeppelin company; on the other, it sought to tame overly enthusiastic requests from Germans abroad for "Zeppelin attention." For example, the German President's Office had expressed considerable wariness over the *Chicago Abendpost*'s request for official greetings to be published after ZR 3's flight: nationalist messages accompanying a reparations machine could indeed cause discomfort in America. By generalizing the message and emphasizing peaceful matters, however, the

German government was able to sidestep the issue and obtain a positive reaction.[64]

The resurrection of the Zeppelin had begun, but the process had the touch of the sorcerer's apprentice about it. Maintaining and channeling the enthusiasm proved to be an enormous task. This was the burden that Hugo Eckener now shouldered, with the help of the German government.

Nostalgia and Fund-raising

The significance of the cultural pride surrounding the airship became clear, in Friedrichshafen as well as in Berlin, as soon as ZR 3 first flew, but there were no illusions that this would secure the Zeppelin company's economic future. Restrictions on the building of airships and the overall economic situation forced the firm to rely on a series of makeshift solutions until funding for a new German dirigible could be found. The key figure in this effort was Hugo Eckener.

The flight of ZR 3 not only brought the airship back into the collective consciousness; it also contributed to the emergence of a new personality in the Zeppelin company who began single-handedly to do everything from managing the firm to leading the flight across the Atlantic. Hugo Eckener, who had been with the company since 1908, now stepped into the limelight and began to displace Alfred Colsman, who had been his superior years earlier.[65] Soon after his return from the United States, Eckener was invited to give various presentations at home and abroad, receiving honors and meeting with high officials wherever he went, all the while promoting the airship.[66] His success in drumming up business and subsidies for the Zeppelin company depended heavily on his genius for public relations. He contributed in large measure to strengthening the image of the airship as a wonder machine, but it took him years to build up the momentum of the airship's popularity.

In 1925 the issue of destroying the airship works dominated the Zeppelin company's concerns. The threat could be removed only by proving that an airship could achieve a peaceful role. One such opportunity offered itself in the possibility of building a machine to explore the Arctic Circle (a project described in more detail in the next chapter). Securing the funds for such an undertaking, however, was another matter. One source of funding was what came to be called the "Zeppelin-Eckener *Spende*" (fund-raising drive).

The *Spende*'s launch, officially intended to commemorate the jubilee of the first Zeppelin airship, came in late August 1925. The organizers of the drive sent out a call for donations in hopes of exploiting the popularity of the Zeppelin story. Eckener later claimed that he understood the risks of trying to resurrect the earlier wave of spontaneous enthusiasm but felt that it was his only hope of saving the company.[67] In any event, the drive represented a novel way of manipulating mass culture. The launch date, August 20, matched no anniversary in Zeppelin history, but it became the occasion for a large celebration in Friedrichshafen.[68] Many celebrities attended, while others sent public telegrams of congratulations and good wishes for the success of the "cultural endeavor."[69]

Getting the fund-raiser under way had required considerable negotiations with German regional and national authorities. Originally presented as a drive to build a "North Pole Zeppelin," the Zeppelin-Eckener *Spende* relied heavily on official endorsements from figures including writers, politicians, and academics.[70] The literary pieces published on the theme of the *Spende* echoed the countless poems and other stories written back in 1908 to honor Count Zeppelin. This time, however, many of the pieces had to be solicited. This does not mean that the *Spende* was an open-and-shut case of public manipulation, for many of the writers offered their work pro bono.[71] Those who contributed their work to the Zeppelin cause were generally conservatives. Many were figures who had gained fame in imperial days but continued to thrive in Weimar times, either as popular novelists or as playwrights, grudgingly supporting the republic as a pragmatic necessity, as Thomas Mann had done.[72] Behind the solicited works of the professionals came the outpourings of the general public. These started as only a trickle, but the flow continued for the remaining years of the Weimar Republic. Although most were perfectly forgettable, they all sought to convey the sublime enthusiasm and sense of purpose their authors had felt upon seeing the airship fly again.

The sense of purpose and endeavor the *Spende* sought to instill attracted people from many different walks of life, a fact that should have greatly facilitated the undertaking. However, the apparently simple formalities turned into a bureaucratic nightmare. The strong "scientific and foreign political significance" of the planned polar enterprise made it almost impossible to discourage,[73] yet its socioeconomic implications worried the Prussian government. Under the Weimar constitution, the states had control of educational and scientific matters. Complaining at having not been consulted, the Minister-President's Office emphasized the ill timing of launching such a fund drive when economic and social conditions remained so precarious. Only in a climate of economic prosperity could the airship's purported benefits really come about.[74]

The Reich government did not object to the *Spende* on economic or scientific grounds, but it did fear that foreign nations would perceive the drive as a warlike gesture.[75] Eckener, with public opinion on his side, rejected these political arguments. He stressed that public donations were the only way to save the airship works and that people in most circles understood the issue in national and cultural rather than military terms. That a company like Zeppelin should have to stoop to begging would only show the world how difficult the situation really was, and how unshakable Germany's determination to affirm its right to participate in "the scientific and cultural field."[76] Eckener made it clear that he would stop at nothing to reach his goals, if necessary by pitting the German people against their government, something his mentor Count Zeppelin had done, too. The Reich made an official statement of support in favor of Zeppelin's survival but postponed the announcement until it could clarify the foreign situation.[77] The parties agreed to emphasize the scientific value of the airship, and Eckener solicited donations from German popular associations and, in the spirit of international cooperation, from sources abroad.[78] Although the moneys obtained from foreign sources were very small,[79] the appeal to Germans abroad reflected the continuing trend of drawing on expatriates to help Germany regain its former greatness. Despite the Reich government's uneasiness regarding such actions, the airship became a means of showing the German flag around the world. The tactic worked, too. By the spring of 1926, thanks also to the Locarno agreement of the preceding fall, the Conference of Ambassadors agreed to lift the ban on airship construction, so that little but resentment of Eckener's "obstinate behavior" stood in the way of the Reich's waiving of its own restrictions.[80]

With the Foreign Office appeased, Eckener began putting less emphasis on the polar-exploration endeavor. He would discuss it only if asked directly and, later, would describe it only as a means toward ends other than exploration. He then exploited the German federal structure as well as contacts at the local level to defeat opposition to the drive. In Cologne, for example, he asked a personal acquaintance, Mayor Konrad Adenauer, to allow the *Spende*. While Prussia toughened its stance against the collection, other states authorized it.[81] To strengthen the Prussian position, Minister-President Otto Braun requested that the Reich release an official communiqué, which he hoped would prevent adverse popular reactions to the government's position while at the same time saving face for Eckener's program.[82] The chancellor was dubious. He recognized the power of public opinion, especially where the airship was concerned.

An official rejection of the *Spende* might actually strengthen the Zeppelin company, which had strong supporters in many circles, including labor unions.[83] To muffle such enthusiasm would in fact promote instability. By the spring of 1926, an annoyed Eckener was arguing that it was up to the people to decide whether they were willing to give 7 million marks for a new airship.[84] Braun relented and authorized collections to start, except in the state's public schools; by summer, the fund-raising was under way.

The German population's awareness of the airship issue, and thus of the *Spende*, was increased by the publication of many works celebrating Count Zeppelin and his machines. The Zeppelin company itself sponsored a sumptuous book filled with the reminiscences of men who had known the count well. The count portrayed in these recollections was a family man whose deeply religious upbringing had given him the patience to endure the stormy years preceding his sudden success in bringing a special gift to Germany and the world; no mention was made of his aggressive public behavior during World War I. The object of the book was less to indulge in nostalgia for its own sake than to invoke the present and future benefits of Zeppelin's legacy. Eckener himself had the final word in the book, in a concluding piece that called for transforming this cultural endeavor (*Kulturwerk*) into a tool of trade.[85] Other books chose to present the count as a supreme technician and entrepreneur in the tradition of Alfred Krupp, Henry Ford, and Alfred Nobel. One theme was common to almost all these presentations, however: the count's accomplishment was the product not just of hard work but of inspiration and even a touch of genius. A man of Christian upbringing trained in the humanities could make as great a contribution to technology and commerce as could an engineer such as Werner von Siemens.[86] Technology was thus an inherent part of culture—indeed, something beautiful. And to see something beautiful in Germany's skies again, the message of the fund-raiser went, it was essential to contribute funds.

Despite the continued popularity of the airship as a symbol of technical and cultural achievement, the Zeppelin-Eckener *Spende* raised only 2.5 million of the anticipated 7 million marks. There was no single reason for the shortfall. Some of it could be accounted for by the management costs of the drive, which included the manufacture of items for sale. But it was also clear that the initiators of the drive and the various government offices involved had simply overestimated public enthusiasm for the project.[87] People were less taken with this orchestrated fund-raiser than they had been with the spontaneous one of 1908; business investors also stayed away. Trying to understand the reasons for the shortfall, Zeppelin engineer Wilhelm Dörr speculated that even though the Zeppelin company was a "representative of national interests . . . German business is still too weak to support such a huge project."[88] He added, however, that another problem also impeded Zeppelin's expansion plans: German airmen, no matter what kind of machine they flew, were considered circus clowns, good only for entertaining the occasional Sunday crowd. This meant that Germany, the "native land of the rigid airship," was going to lose out to the business-oriented British when they completed their own transport dirigibles. If the new German airship was built and sold abroad, following the path of ZR 3, then "we would have nothing left."[89]

Still, the fund-raiser was considered a success and paved the way for proving the Zeppelin mettle through deeds rather than words. Construction of the new LZ 127, like that of ZR 3 some four years earlier, was a highly publicized affair,[90] but the financial arrangements remained shielded. While the government contributed some 1 million marks toward the project, another 3.5 million came from Zeppelin subsidiary companies that Colsman had set up.[91] Eckener's gamble, then, required the very "pots and pans" of postwar conversion that he had opposed. The press reported at length on the technical aspects of the new machine, marveling at its size and expected power.[92] There was, however, one obvious difference between ZR 3 and the new apparition: unlike its predecessor, the new dirigible would remain German property, "a piece of ourselves in the sky" rather than a "cultural item for other nations."[93] In fact, since the British were also in the process of building airships, the aeronaut Gustav Milarch predicted a peaceful competition for a "blue ribbon of the air," reminiscent of those awarded for speed records to seafaring ships.[94] The sense of optimism that had accompanied the building of ZR 3 was even stronger now, since economic conditions had improved notably and the destruction of the Zeppelin factory had ceased to be an issue.

Christened *Graf Zeppelin* in July 1928 while still undergoing factory checks, LZ 127 did not differ markedly from ZR 3. The need to increase its lift capacity accounted for a difference in shape, but in other technical and external respects the machine was a blend of standard features and tested innovations. At 236 meters, the *Graf* was one of the longest airships ever built, but the size was dictated by the length of the assembly hangar, not by special aerodynamic calculations. The point of the design—to prove the feasibility of long-distance airship travel—allowed for the installation of a few novel features. The passenger cabin was placed just behind the control room (as on ZR 3 and LZ 120),

The *Graf Zeppelin's* lounge. The lounge also served as a dining room, and unlike the arrangement on the *Hindenburg,* no smoking was allowed. Passengers described the floating sensation as akin to that of an ocean liner, but "without the sea-sickness." In storms, though, the lack of floor bolts sent furniture and occupants flying in all directions. UTD, Rosendahl papers, box 14, folder 7

but this time small rooms with beds were set up, each including a wash basin. The kitchen received a full-size oven and refrigerator, in expectation of flights that would last several days.

The real innovation, however, came in the use of Blau gas instead of gasoline as an engine propellant. All dirigibles encountered the same problem in flight: as the engines consumed propellant, weight decreased and the machine would rise. To maintain a stable cruising altitude, the crew would vent hydrogen through specially designed traps, but this was risky because weather conditions might require the ship to rise again. The practice was also very expensive. If, on the other hand, the propellant could be replaced with ballast as it was spent, then the crew would have to vent hydrogen only in an emergency or during landing. In theory, using a gas as both lift element and propellant might offer a partial solution. In the nineteenth century, the Austrian pioneer Paul Hänlein had experimented with this process by using coal gas in his airship. This would not work with hydrogen, since at the time there was no known way of controlling the gas's extreme volatility. Over at Zeppelin's, the physicist Eberhard Lempertz, who worked as the company's meteorologist, suggested an alternative: use two sets of gas cells, one containing hydrogen and the other, a propellant gas with a density close to that of air. The solution was Blau gas (named for its inventor, Dr. Hermann Blau), only some 9 percent heavier than air.[95] As the fuel was consumed, the cells decreased in volume and air took up the space. The engines for LZ 127 were therefore modified to accept Blau gas but also to run on kerosene for landing and takeoff maneuvers (Blau gas was later replaced with a propane mixture, judged to be more efficient).

Once the technical troubles were ironed out, test flights of the new machine took it over Germany and parts of Switzerland. Unexpected "incursions" over the occupied Rhineland and the eastern coast of England elicited an enthusiastic response from the German population and unhappy comments from Allied authorities.[96]

The long-distance test flight over southern Germany on 28 September 1928 linked the airship with cultural and social life. "He flies," commented a weekly paper, choosing to masculinize the gender-neutral machine: "With him rises a piece of ourselves; for twenty years, the Zeppelin has been in the heart of Germans; the millions of marks gathered after Echterdingen proved it, as do the other millions of marks poor Germany sacrificed for him."[97] Foreign Minister Stresemann, whose relationship with Eckener was somewhat strained, wished the upcoming flight good luck.[98] On this occasion the Reichstag president Paul Löbe came on board, along with Oscar von Miller, the founder and director of the Deutsches Museum, and the painter Ludwig Dettmann, who immortalized the event.[99] The flight also became the occasion for one of the earliest public radio addresses from the air, in which Löbe, comparing the flight to a comfortable train ride, admired the beautiful view of the landscape afforded by the airship, and hoped that the people he saw running into their homes did so to fetch their children and show them "man's wonder work." After Löbe's remarks, von Miller came on the air and proceeded to describe the Deutsches Museum's collection of airship memorabilia.[100] This combination of radio and the airship came to represent the peak of modernity, demystifying the world and broadening the horizons of knowledge and of mass culture.[101] Only the Communist press openly opposed the *Graf Zeppelin*'s presence, comparing the engine noise to the rumblings of a "reactionary guard" announcing a new wave of German imperialism, starting with a "business tour of the nation."[102]

The new airship's entertainment value also increased its importance in

Weimar popular culture. Novelists exploited its modernity, depicting, for example, a "flying hotel" dirigible aboard which French agents plotted against German interests as the machine flew around the world.[103] Early versions of today's "techno-thrillers," such works had little literary merit, yet they continued to grace the shelves of public libraries alongside the war literature that had begun to reappear.

A collective memory of the Great War slowly surfaced in the 1920s, at least in the literary world, coinciding with the *Graf Zeppelin*'s early successes. Some accounts of the air war, contrasting experiences at the front with treachery at home, became a means of articulating extreme nationalist views.[104] Although based in fact, such accounts served to create a new, selective memory of the conflict in which duty and sacrifice were the main focus, overshadowing such themes as defeatism and the dangers of technology. Airshipmen's war accounts enjoyed a limited but steady success. They did not portray the war as such, at least not in the way the trench-warfare accounts did. They downplayed the gruesome aspects of death in the air and emphasized the glory of flying for the country, the spirit of camaraderie among the crew, and, of course, the heroic feats of the individual aces.[105] All of this contributed to a sanitized and romanticized view of the war. In one popularized view, German airships—despite their actual limited effectiveness—had helped at the front by keeping some five hundred thousand British air-defense troops at home to deal with the Zeppelin threat.[106] Further glorification revolved around a personality cult. A common theme in accounts of the airship in the Great War was the near-deification of the naval airship commander Peter Strasser, who had died on one of the last missions to bomb London. His unswerving belief in the superiority of the airship as a war weapon—an echo of Count Zeppelin's views—could only end in a "hero's death . . . as a fulfillment of duty."[107] It was sacrifices like this, according to the postwar interpretation, that had permitted the airship's technical progress.[108]

Depictions like these clouded the official message of the airship as an instrument of peace. The glorification of resistance to the Allies had a similar effect. In particular, the so-called Scapa Flow of the Air of summer 1919, when crews of grounded airships followed the example of the German fleet and destroyed their machines rather than hand them over to the Allies, suggested a spirit of sacrifice that would help keep a German symbol alive in memory if not in reality.[109] The retelling of incidents like these reinforced the airship's image as a uniquely German technical success, so special that only Germans could care for it, and imparted to it a quasi-organic, even human, stature. Less than a decade later, the *Graf Zeppelin* would assume a similar status.

Symbol in the Best and Worst of Times

Following its triumphant flight around Germany in September 1928, the *Graf Zeppelin* undertook several more test flights in preparation for its first major task: a transatlantic crossing. Eckener had publicly stressed many times that long-distance airship links were the wave of the future, and he was eager to repeat the ZR 3 triumph from four years before. This time, he hoped, the crossing would be the first of many routine flights. In the meantime, a young American airplane pilot by the name of Charles Lindbergh had completed the first transatlantic solo flight. Other flyers had followed, including one who completed a New York–to–Berlin jump. In each case, however, the public's interest in the events died down fast, and the flyers were often dismissed as "publicity cravers."[110] Yet the rarity of transatlantic flight still made such voyages an object of considerable attention, and an airship "jump" was no exception. In October 1928, evening tabloids followed daytime papers in reporting the latest sightings of and radio contacts with the *Graf.* The airship's final destination was Lakehurst, New Jersey, a naval airship base close to New York City with a field wide enough to accommodate an airship landing and the necessary crew to assist in maneuvering the machine on the ground. ZR 3 had landed there in 1924, and Eckener, who had captained it, also commanded the *Graf Zeppelin* as it made its way to the American East Coast. When the airship landed, an Austrian columnist defended the achievement as greater than Lindbergh's: Eckener's flight was a commercial enterprise, not just a show.[111] This interpretation rather blurred the motivations, since Eckener himself expected the publicity associated with the flight to help fund his endeavors. Two weeks after arriving at Lakehurst, the *Graf* flew back to Friedrichshafen, its return prompting new waves of public rejoicing that echoed the enthusiasm that had surrounded ZR 3's previous success.[112] From now on, the machine itself became as much a symbol of Weimar Germany as earlier Zeppelin models had been of the empire.

The *Graf Zeppelin*'s shape, essentially a function of evolution in airship design, also seemed to reflect the modernist aesthetic of 1920s Germany and America. Although the New Realism was well advanced in the world of art and architecture, in the industrial world, product styling and standardization practices remained in their infancy, despite a move to dispense with ornament and superfluous detail.[113] The *Graf* offered a kind of compromise between the worlds of art and industry, in two senses.

First, the *Graf Zeppelin*, like its two immediate predecessors, had a purity of form compared with earlier airships, with a cabin merging into the fuselage and reshaped engine nacelles. This kind of streamlining epitomized progress and suggested speed, even though velocity had in fact barely increased.[114] The airplane was flying faster than the airship, but to people outside of aviation, the airship "looked the part" better. Airplanes remained sharp-edged, noisy, and shaky, and they stank of gasoline and oil.[115] The airship, more comfortable, suggested a sturdy assuredness as it sailed through the sky, dwarfing the clouds and casting a shadow over entire villages.

Second, the *Graf Zeppelin* was the only one of its kind, and its construction, although requiring industrially produced elements, relied on machinists working like craftsmen. The machine's uniqueness increased its symbolic value. The fact that it was a prototype saddened some, who felt that it was an "aristocratism" in a time of democracy. Others suggested that its uniqueness warranted some peculiar comparisons with human genius. Years later, during a South Atlantic cruise in the "silver fish's belly," the writer Heinrich Jacob marveled at the machine's internal structure in such terms: "What Mozart is among composers, aluminum is . . . among metals. Firm, yet light."[116] But the airship offered more than cultural and moral symbolism; it remained a business proposition, for as Jacob added (perhaps under the influence of Eckener), "It brings the German businessman to Rio in four days, saving him another thirteen."[117]

The media capitalized on the economic element. Promotions using the image of the airship appeared even before the *Graf Zeppelin* had returned home from its first trip to the United States. Advertisements for the latest popular study of the airship, the brand of alcohol consumed on board, and even health pills all vied for readers' attention.[118] A Berlin map manufacturer printed on his globes the route the airship had followed. The *Graf* thus joined the unlikely collection of objects that fascinated Weimar society at its zenith. This fascination extended to fast machines of all kinds—motorcycles, rocket test vehicles, airplanes. An experimental bullet train became known as the "rail-Zeppelin" because its streamlining was reminiscent of the airship's.[119] What differentiated the airship from other mass-culture technologies, however, was the apparent inclusiveness of the Zeppelin as a symbol. Whereas other manifestations of German mass culture—film, for example—could be instrumentalized

Where size matters. Advertising gimmicks pushed the uses of the Zeppelin motif into new realms. Here, the Mira company claims that if all the cigarettes it has produced so far were laid end to end, they would stretch from New York to Berlin, a 6,000-kilometer distance as flown by an airship. During the Apollo moon landings, an American fast-food chain used the same analogy: a stack of all the hamburgers it had sold would reach to the moon and back. *Kölnische Illustrierte Zeitung*, via ASTRA

The *Graf Zeppelin* is brought into its shed. This operation required as many as three hundred men. In Germany, the event would attract volunteers who vied for the honor of participating. Occasionally, schools relied on connections with Zeppelin officials to initiate special class trips to the Zeppelin base and involve teenagers in what became a kind of patriotic ritual. UTD, Rosendahl papers, box 17, folder 8

to serve the purpose of Germanness, there was no need to do so with the airship, for it *was* in itself Germanness. Indeed, popular works that jointly glorified German modernity and tradition depicted the airship alongside cranes on the one hand and images of the traditional German countryside on the other.[120]

The Zeppelin fascinated children. School groups vied for the chance to visit Friedrichshafen and even help pull the giant out of its shed.[121] Such opportunities were rare, however; children were more likely to encounter airships in books than in a hangar. Erika Mann, daughter of the writer Thomas, chose the airship as the focus of her 1932 work for youth. In another book, Eckener himself was the hero of the airship history, held up as "proof that courage and self-control lead to the highest success."[122]

This reference to hard work was but one of many in such popular works that suggested a return to the values Count Zeppelin had advocated: a positive outlook expressed in the form of "unflagging work."[123] Writers continued to emphasize the Germanness of the airship as well. Unlike other technologies, it did not have to be reconciled with the German spirit (*Geist*); it was already part of that spirit, and thus in harmony with the people.[124] The airship also had a remarkable potential for peaceful purposes. Measuring the cultural propaganda value of the airship, the *Kölnische Zeitung* suggested that "the [Balkan] flight [in the spring of 1929] . . . was for us Germans, as well as foreign lands, . . . a probation trip, the cultural political value of which is hard to overestimate." The *Deutsche Allgemeine Zeitung* summed up the dirigible's emblematic power thus: "The Zeppelin [is] our best German ambassador."[125] The Reichsbank capitalized on this notion by minting Zeppelin commemorative coins for sale at home and abroad, hoping to foster goodwill and an understanding of Germany in nations where the Zeppelin itself had not flown.[126] Conscious of the machine's cultural impact, the government sponsored flights and invited dignitaries and journalists to attend, thus acknowledging the close connection between politics and culture, even in a technological setting. "Zeppelinism" had evolved, but it remained almost as strong as the Wilhelminian brand in that it was now meant to reach beyond German borders.

The *Graf Zeppelin*'s 1929 trip around the world was a case in point. Combining business with nationalism, Eckener obtained the backing of American press magnate William Randolf Hearst, who demanded—in addition to exclusive coverage rights—that the trip begin on American soil. The airship first flew from New York to Friedrichshafen, where another "official" world tour departure took place on August 15, so as to satisfy German sensitivities. Four days later, the *Graf* reached Tokyo, prompting stunning displays of enthusiasm for all things German, and even a rare imperial reception.[127]

Although two of the *Graf Zeppelin*'s four destinations were American cities (Los Angeles and New York), some of the more enthusiastic accounts of its trip around the world actually appeared in the form of adventure books published abroad, even in nations the airship had not overflown.[128] Advertising also capitalized once more on the success, particularly advertising for German technology. From Leica cameras to Krupp crankshafts, anything associated with the trip was offered for sale, thus indirectly confirming Eckener's claim that the airship would help world trade. Not only had the machine drawn continents closer together, but the "empire of speed" in which it sailed had turned "Japanese and Chinese into Europeans."[129]

The world flight of the *Graf Zeppelin*, completed in twelve days, proved to

Zeppelin marketing: detergent. As an element of modern life, the airship became associated with both consumerism and the idea of speed. Presumably, Hoffmann's detergent would make quick work of household chores. This tin advertising sign was placed outside shops selling household goods, ca. 1925. Author's collection be of tremendous value in impressing other nations with the values of a democratic German republic. The country's finances appeared stable, thanks to heavy American investment in the German economy. The world agricultural depression was wreaking havoc on German farms, but a matter of greater international concern was Germany's protests over the Young plan, which set the rate of payment of the reparations mandated by the Versailles treaty. The Zeppelin may have been a machine for all Germans, but it could neither solve the ongoing problem of Germany's punishment—although Eckener had hoped that the airship would help communicate goodwill—nor reverse the signs of economic trouble ahead.

Indeed, the sense of national kinship and other sentiments that surrounded the airship could at times be a source of misunderstanding. For example, honoring the *Graf Zeppelin* on its return from its world flight would no doubt boost the government's popularity at home, but what kind of honors to extend became an issue, especially with regard to the "flagging" of official buildings. Without consulting his superior, Transportation Minister Adam Stegerwald and his assistant for aviation matters, Ernst Brandenburg, called for all government buildings to be draped in the national colors, thus angering several colleagues and superiors in other federal and Land ministries.[130] Not only was there a problem of bad timing—the flagging might coincide with the official commemoration of the Allied occupation of German regions—but to have central government services acknowledge the airship's achievement while local administrations ignored it might send a message of national disunity. On the other hand, students in Prussia were let out of school for the occasion, without any federal mandate ordering this action. In the end, the Prussian school decision was passed on to other states for consideration, and flagging took place only in states where the Land-government issued such orders, such as Württemberg.[131] This apparently insignificant incident nevertheless shows that despite its obvious value as a cultural icon, the airship did not quite fit into the traditional realm of official public celebration and commemoration. Celebrating the airship was not an entirely straightforward matter, perhaps because of the political circumstances that had made its reappearance so laborious. Nonetheless, the positive modernist associations of the airship remained a strong point at all levels of its influence.

The *Graf Zeppelin* became the cherished vision of Germans in the late Weimar era. Its greatest impact on German public opinion may have been its visits to the Rhineland in 1930. The end of the Allied occupation of the area that summer became the occasion for a multitude of celebrations, including aeronautical meetings. The *Graf* landed twice on Rhenish soil in three months, creating a sense of elation that the press agreed had rarely been felt before. "Like a vision, a silver tale from the Thousand and One Nights . . . *He* was there!" exclaimed a Bonn columnist in April of that year.[132] The airship's visit was as much a way of rejoicing over national liberation as of claiming geographical importance. Bonn, a sleepy provincial town, could claim to have beaten neighboring cosmopolitan Cologne to this one-of-a-kind "play," attracting thousands of spectators from all over the Land on Easter Sunday, "for in these times, not to have seen a Zeppelin is to profess cultural ignorance."[133] Similar reactions accompanied the official "Rhineland Freedom Flight" a few months later.

The *Graf Zeppelin* returned to the Rhineland three months later, this time as part of an official celebration of the recovered right to fly to the region, "a popular feast in the spirit of the century of technology."[134] A carefully choreographed ballet of small airplanes opened the show, followed by a huge Junkers G-38 carrying government and state officials. The display culminated in the arrival of the airship.[135] "Crew calls rang over the field. The airship landing team stirred. The crowd waited excitedly. They gazed jubilantly at the nacelle windows, from which the passengers returned their greetings. The air giant's nose sank to the ground, the holding ropes unfolded, the proud eagle of the skies lowered his wings and let himself down closer to the ground. Hundreds of hands seized the nacelle's body." The airship had come home. Like a good-natured giant out of a children's tale, epitomizing the magical side of technology, it meant no harm and could bring only joy in times both good and bad. An emblematic liberator, humanized and lionized, the rumbling silvery device left "an unforgettable memory in the heart of all Cologne people."[136]

The question remained, however, whether the airship was still a symbol of universal Germanness. Although it brought pride to German communities abroad, its impact in Austria differed from what had occurred during Count Zeppelin's 1913 visit to what was then a dual empire. Now, as the first republic struggled to its feet, Austria was dealing with its own identity crises. The *Anschluss* question resurfaced forcefully as one of the revisionist goals on both sides of the border.[137] In contrast to the Hitler years, in Weimar Germany parties of the center and the left were initially the most active in advocating a union between Germany and Austria as a way of consolidating the fledgling Weimar Republic. The writer Hugo von Hoffmannsthal, who years earlier had marveled at the visiting *Sachsen,* wrote pessimistically of the future of an Austrian national identity.[138] In view of these issues, Zeppelin visits to Austria might have been a welcome sign to Austrians longing for political and cultural stability.

The *Graf Zeppelin's* flights also took it to Central and Eastern Europe, but few of these trips ever included Vienna on the way, possibly because of fears in both Vienna and Berlin that such a flight might stir up nationalist feelings and anger the Allied nations. A cursory flyby without a stopover in the spring of 1929 had left the Viennese public feeling angry and shortchanged, but the Austrian government was understanding about it; the transport minister wrote a note thanking the German ambassador for deepening Austrians' "feeling of linkage to the great German brother people" and furthering the conquest of the air.[139] While couched in diplomatic language, the note bore unmistakable testimony to the official perception of kinship between the two nations, and to the young Austrian republic's difficulties in finding its own kind of nationalism.

The reaction on the German side to the hasty nature of the Vienna overflight also confirmed the perceived sense of kinship. Reminding his readers that the "second greatest German city" should not be treated as a simple relay station, one columnist suggested that Vienna would have liked to rejoice over the success of German achievement.[140] Echoing such views in a peculiar display of pan-German feeling, the Catholic newspaper *Germania* pointed out that even without *Anschluss,* the Austrians had displayed surprising solidarity by rejoicing over ZR 3's flight to America in 1924 and by contributing to the Zeppelin-Eckener *Spende.* The least that Germany could have done in return to maintain good relations and save years of "petty diplomatic work" was to fly to the Austrian capital to display a "high style" example of Germanness.[141]

With Austrian sponsorship, the *Graf Zeppelin* finally descended over Vienna one July morning in 1931. The city and country had changed considerably since the visit of the *Sachsen* eighteen years earlier, but the airship remained a symbol of German culture and technology throughout the world. Unlike the last time, few commentators now mentioned the Austrian pioneer David Schwartz, despite a controversy over his role in inspiring Count Zeppelin's own designs.[142] This airship visit would serve as grounds for celebration, not debate. Upon landing, the *Graf* was accorded a reception worthy of a head of state: both the president and the chancellor were on hand to greet Hugo Eckener, and a series of speeches emphasized the strong link between the Austrian and German people, carefully skirting the uncomfortable question of joint Germanness.[143] Some newspapers abandoned neutral ground and raised the issue by comparing the visits of the *Graf* and the *Sachsen*.[144] Some felt that unlike the previous visit, which had taken place for "the old man in Schönbrunn" (Emperor Franz-Josef), this one was an event in which all Viennese could rejoice, for they were part of the spirit that had built this astonishing object. The feeling of belonging, however, was limited to cultural kinship and did not extend to a nationalist one, for as a *Tag* article concluded, "'Graf Zeppelin,' we Viennese, we Austrians salute you!"[145] The Communist *Rote Fahne,* predictably more rabid, described the "national rumble" of the machine as symptomatic of a warlike tendency. It did add, however (perhaps as a face-saving measure to its readership), that those who had come to see the airship were for the most part present to admire a masterpiece of human achievement: attendance reached over one hundred thousand, including many working-class people, according to police reports.[146]

Many of the *Graf Zeppelin*'s successes accumulated at the end of the Weimar Republic's stabilization era, when Foreign Minister Stresemann was seeking to mend Germany's economic and political status through rapprochement with France and Great Britain. The overall perception of the airship remained that it was a reflection of German success in all fields, just like the transatlantic flight of the Junkers *Bremen* aircraft in 1928 or the Dornier DO-X giant seaplane's carrying 169 passengers on a test flight. Eckener, for his part, tried to dispel this assumption, asserting that only a truly free Germany could contribute to human development. Otherwise, such German achievements as the ship *Bremen,* which set certain speed records, and the *Graf*, described in the German press as holding the "Blue Ribbon of the globe,"[147] were but products "of the generation bred before the war" and testified only to a strong technical heritage;[148] modernity could not fully mature without a strong political agenda.

Yet strong technology in itself was the very argument that Eckener had used to bring back the idea of the airship. Even before the *Graf Zeppelin*'s construction in 1927, business leaders had renewed the call for ways of making air transport profitable: technology existed not for its own sake but for the sake of helping business and thus the nation in the wider scheme of things. At the time, airlines flew only thanks to heavy government subsidies, and the same was true for the airship. Eckener was unhappy about the situation, but for lack of other options he constantly sought to justify the government's financial support.[149] In most of the interviews he gave he emphasized the long-term investment the airship constituted.[150] He also forwarded to the authorities letters from Germans enthusiastic about the machine's achievements. Such compliments were all the more important when they came from Germans abroad and yet echoed the sentiment at home. "You have rendered German industry an inestimable service," wrote one businessman returning from two months in the United States; since the *Graf*'s world flight, he added, "every German in the United States is again proud of his Germanness and is recognized as an equal by the Americans."[151]

The airship's strong association with *Americanism* could not go unnoticed. A German diplomatic report stressed the importance the American press assigned to any airship activities and the positive impact such reporting had on the public. The landing and transfer of passengers had become smoother and more efficient, despite the masses of people coming to Lakehurst to witness the *Graf*.[152] Although airship technology was becoming routine, it continued to exert an almost magical attraction, which might work to the advantage of business interests. An informal survey of the German-American Chamber of Commerce in New York reported that all the trade associations surveyed considered airship flights an excellent tool for promoting a climate conducive to business. The reasons, however, varied considerably. To one group, the crucial feature was the airship's time-saving ability—the basis of all successful business deals. To another, it was the marvelous technological know-how that had built the *Graf Zeppelin,* which would inspire trust in other German manufactured products. Yet another association claimed that it was simple admiration of the airship, which "helped break down sales resistance."[153] All the groups surveyed emphasized the goodwill that the airship inspired. Nevertheless, their responses all had a noncommittal tone, reflecting the poor economic climate at the time and an unwillingness to invest in a technology that, for all its promises, remained experimental.

Yet Hugo Eckener chose to emphasize these very promises. Bringing Germany to the level of other nations economically and politically had long been a goal Eckener shared with Foreign Minister Stresemann. Stresemann had hoped to achieve revisionist goals through rapprochement, but his death in October 1929, together with the Wall Street crash, became grounds for a reversal of views on the matter. Calls for the cancellation of reparation payments became legion. Eckener was a strong voice in the shift toward a more aggressive revisionism, using his highly visible position to attack the Young plan and demand its abolition because it hindered his country's progress, and thus the world's.[154] The internationalist position in which peace was the main message shifted to one in which Germany had its role to play, with the help of aviation and the airship.[155]

The Wall Street crash did not hurt the Zeppelin company right away, but it nevertheless prevented expansion plans and joint ventures from going through. As foreign investments disappeared, the airship persisted as a solitary example of Germany's technical achievements, a lone symbol of patriotic duty. Eckener used this fact to argue that the machine would contribute to help German business in the long run. Throughout the stabilization and decline periods of the Weimar era, he became as much a fixture of German news as his machines were, traveling tirelessly to give speeches the world over on the benefits of the Zeppelin. He remained a shrewd businessman. Even when the economic crisis hit Germany, he continued to sell his idea of making *Zeppelin* synonymous with *business*. In so doing, he also fed the popular image of aviation that pervaded the Western world, whereby geopolitics was to extend into the aerial realm, granting new avenues of expansion to any country in the lead.[156]

Several Zeppelin associates privately voiced concern about the role Eckener had assumed. Colsman had retired, and if Eckener, now sixty-one, were to die, the whole enterprise would collapse, for it all seemed to rest on his flair for public relations. None of these concerns became public, however, and Eckener continued exploiting his own popularity as a means to advertise the airship. Using evidence gleaned from the *Graf Zeppelin*'s previous flights, he argued that regular airship traffic with the American continent would bring stability and business opportunities to German companies. The Transportation Ministry appeared willing to believe his claims, for it sponsored several of his talks to the public and to captains of industry. Many commentators were convinced that the airship as a "business factor" would strengthen Germany's international reputation, even in troubled times, a fact that made subsidies a worthwhile sacrifice.[157] Unfortunately, the economic situation remained dire, and the measures that were taken to keep the Zeppelin in the air were more an emergency response to pleas from the company than a result of careful planning.[158]

From almost two hundred flights in the period 1930–31, the *Graf Zeppelin*'s activities slowed to less than sixty flights in 1932. Economic conditions did not allow much more than the planned transatlantic flights as well as a few charters to pump money into a company already surviving on subsidies. Yet the airship symbol remained very much alive in the declining years of Weimar. Its value as an entertainment item during difficult times was high. In the sunny days of Weimar, air shows had proven quite popular. War aces and daredevils drew huge crowds, but the airship, though incapable of anything more than a slow flyby, was an even bigger attraction. Whereas pilots like Ernst Udet could draw some sixty thousand people to an air show, an airship guaranteed almost double that number, even if it was the only exhibit.[159]

Prospects like these were of great interest to cities faced with dwindling resources and the threat of social unrest. The costs involved in bringing the airship to a specially prepared fairground appeared prohibitive at first, since the Zeppelin company required an average of 10,000 marks per charter flight.[160] But even with the infrastructure costs, benefits would come at several levels, including the basic one of popular entertainment. As an official in the town of Bielefeld put it, "The Zeppelin days in Bonn and Münster have resulted in such huge incomes that the chartering aviation associations were left with considerable benefits."[161] The attraction was such that according to the same official, group trips from Holland were already expected simply on the basis of the rumor of a landing. The sense of local pride occasioned by a Zeppelin landing was accentuated in hard times. The town of Meiningen in Thuringia, for example, scraped up all available funds to charter an airship landing in the fall of 1931. When the machine landed, a brawl almost ensued between local spectators and neighboring Prussians who had crossed the land border to witness the arrival of "their" Zeppelin.[162]

Even though the New Realism was by now giving way to cultural pessimism and political polarization, the airship remained a national symbol of hope, thanks to the sense of the dynamic sublime that it offered and that crowds insisted on seeing.[163] The conservative DNVP party even used this aspect of the airship theme on one of its campaign posters.[164] The DNVP was not the only political body that tried to use the airship or its main advocate, Eckener, to rescue a faltering political system. Such a uniting factor might prove the ideal solution in the face of extremism: a conservative voice that proclaimed the ideals of the republic.

The worsening economic situation further strained the Weimar political fabric. By the spring of 1932 it appeared that Adolf Hitler was challenging Paul von Hindenburg for the presidency. Eckener, who had always shunned open involvement in politics, declined Chancellor Heinrich Brüning's request to make the *Graf Zeppelin* available for propaganda flights in support of Hindenburg, although he agreed to endorse the incumbent in a radio speech.[165]

When it appeared that Hindenburg might not run again, however, Eckener, with the encouragement of friends, sought to run against Hitler himself. Eckener was one of the highest-profile figures in Germany, and his chances of winning the elections were considered to be quite good, since he also had the support of the Social Democrats despite his conservative sympathies.[166] However, as soon as Hindenburg changed his mind, Eckener withdrew, leaving the political scene but having unwittingly incurred the private wrath of Hitler. The move would hurt him in the years to come, but the episode also illustrates how his actions in behalf of the airship had made him as famous as the machine. The charisma he had shown from the start in developing and defending his idea of technology and its role in business almost paid off, paradoxically, in politics, and would slow the Nazis down in their use of the airship.

Technology assumed new roles in the Weimar Republic. To some commentators, it was a force that would expand industrial, and thus capitalist, success by "solving the greatest problems."[167] Others, however, distrusted its potential, for it had introduced a new and debasing kind of warfare. When it came to the Zeppelin, however, both sides tended to suspend their antagonisms and agree on the symbolic importance of this machine, which blended so many multiple messages. To those hoping for reconciliation between Germany and the rest of the world, the airship was a token of friendship to be sent ahead. To those professing a stronger nationalism, however, it offered a beacon in the night of Versailles.

German culture and society had embraced the image of the airship with few misgivings. To put it in terms of developments in the art world, in the airship there seemed to be no gap between the subject-matter and the art-object. The airship was both simultaneously, whether in the ongoing excitement it caused or in the benefits it might bring. On the one hand, it fit Siegfried Krakauer's characterization of the technology world in which machine worship and the affirmation of the mass person were at work.[168] On the other, it attracted the attention and support of the intellectual and bourgeois elite in a manner remarkably reminiscent of the reception the airship had received in 1908.

In terms of the technology itself, the airship came to maturity in Weimar Germany. Lessons gleaned from the war were incorporated into new designs, but an evaluation of the new items reveals them to have been more a matter of fine-tuning than of outright innovation. Improvements were seen in speed, hydrogen use, and lift capacity, but the problems in these areas had still not been fully solved. The eventual construction of the giant *Hindenburg* in Nazi times appeared to represent progress, but many of its features in fact dated back to the Weimar era.

The role the Zeppelin assumed during the stabilization and decline of the Weimar Republic clearly shows its importance as an instrument of popular culture and as a political and business tool applied to the rehabilitation of Germany. "Eckener's technical feats drew attention and sympathy to the German nation," noted a prominent Social Democrat.[169] Tensions in the cultural and political fabric of the era apparently disappeared where the airship was concerned. The airship was not simply a symbolic rallying point for the overturning of Versailles but also tangible proof of a German legacy from the imperial era.[170] ZR 3 and later the *Graf Zeppelin* thus represented national strength in a weary yet peaceful world. Commercial gain was the professed goal of the airship, but that goal often came across as a means to support the ends of the symbol, rather than the reverse. Alfred Colsman himself acknowledged that before the issue of air transport profitability could be raised, "Friedrichshafen had to be seen as a duty, not a business."[171] Hugo Eckener's contribution to this endeavor was a brilliant public-relations campaign that presented culture and business not as opposites but as complements in the wider scheme of airship technology.[172]

The Zeppelin offered, in effect, a nostalgic trip with utopian overtones. Much as France had sought in its *années folles* to recreate parts of its *belle époque*, so Weimar Germany used the airship as a means to remember an aging Count Zeppelin who had given united Germany a happy sign of its youthful promise. Expressions of pride, however, also reflected tensions in the social fabric, which opponents of Weimar democracy were eager to exploit. Thus, the airship had to negotiate a fine line between nationalist revenge and modernist patriotic pride, relying on a peculiar "structure of feeling" to appeal to the masses.[173] The extreme political manipulation of the symbol was yet to come, as the Zeppelin company proceeded half-heartedly to invest meager funds in the construction of a new airship in 1932. Whereas the completion of its predecessor had required minor compromises with the Weimar Reich government, Eckener would soon discover how difficult it was to make similar arrangements with the National Socialists.

CHAPTER 5

Ideologies of Science
and Adventure
The Arctic Airship

OW SIMILAR WERE THE ZEPPELIN EXPERI-
ences of the imperial and Weimar periods? How had
the technological symbol evolved? Something had changed, now that the inventor was no longer alive; it had taken a considerable amount of time to rekin-
dle public interest in the airship after the war. There was obviously still a Zeppelin myth at work, conceived as a story that maintained and justified present
actions.[1] But Hugo Eckener, the protector of the Zeppelin legacy, needed more
than stories to warrant the millions of marks that supported the firm's activities. He set specific goals to keep the company afloat and used various means
to justify them, emphasizing economic reconstruction and the role the airship
would play in it. He also offered other ideas to match new moods. These included the planned flight of a dirigible to the North Pole, a project inspired
by a scientific goal of the imperial era. In promoting this project, Eckener played
up the rhetoric of science and turned it into a marketing tool; science "had to be
served, but it could serve as well."[2] While Eckener grudgingly agreed to set some

scientific goals for such an airship expedition, it was only to showcase the technology itself.

The story that culminated in the celebrated 1931 flight of the *Graf Zeppelin* to the polar circle originated long before the foundation of the republic. German efforts to investigate the polar region by air represented a confluence of scientific progress, economic self-interest, political dealings, and popular fascination with what was then one of the last unexplored regions of the globe. By charting the evolution of this idea, one can understand more clearly the airship's development as a symbol between the imperial and Weimar periods.

Polar Dreams

The Arctic region became the focus of geographical interest beginning in the sixteenth century, when merchant expeditions attempted to find the so-called Northwest Passage that would offer a quicker route to the Eastern Hemisphere. In the nineteenth century, Western powers stepped in, backing one exploration team after another. By then, reaching the North Pole had become the main goal of northern expeditions, with the goal finally being achieved by Robert E. Peary in 1909.

Despite the end of the Arctic race, enthusiasm for similar projects persisted. Ballooning added a new dimension by theoretically making possible long-distance flights over both oceanic and mountainous areas.[3] Among the visionaries who undertook such expeditions, the Swede S. A. Andrée, with two assistants, set out in 1897 on board the balloon *Eagle,* hoping to reach the North Pole. The aircraft crash-landed on White Island, and the crew perished the following winter. Their remains along with their journals were not found until 1930.[4] Like so many other mysteries of the frozen north, however, the disappearance of the members of the Swedish expedition only increased popular fascination with the "white desert." Many advocates of further exploration believed that despite the failure of their expedition, Andrée and his team had proven that an exploration of the Arctic by air was possible.[5] By the time Peary reached the Pole, Andrée's would-be followers had tried at least fifteen times, unsuccessfully, to fly to the North Pole from the Spitzbergen archipelago in the Barents Sea.[6]

Dreams of achieving the seemingly impossible were part of a larger mood of utopian optimism, tied to the belief that science, if applied diligently, would serve humanity and its problems. New theories rocked the international scientific community of the day, from Einstein's theory of relativity to Alfred Wegener's continental-drift hypothesis. But science at the beginning of the twentieth century included far more notions of knowledge than it does now. To explore in any way at all was to practice science, and a common adventurer using machinery to reach his goal might be taken as seriously as any theoretical physicist. *Homo technologicus* might become the proverbial Dr. Livingstone, carving out a new place in the annals of exploration history.[7] Political considerations affected science, too. The age of imperialism had almost run its course. The great powers' "scramble for Africa" and their claims to possessions in the Far East had left little land to acquire. Although their geopolitical potential remained unknown, both polar regions were among the last of the *terrae incognitae* that had peppered ancient maps. To chart these would both demonstrate the "scientific" advancement of Western civilization and affirm individual nations' geopolitical and political-economic "greatness."[8]

Governmental support for geographical exploration could prove lucrative for a variety of reasons. In the case of Wilhelminian Germany, assisting science also served to consolidate the imperial structure. *Verreichlichung* (federalization) occurred wherever the government thought scientific undertakings might be related to the cultural interests of Germany.[9] Economics also played an increasing role in the decision to support certain endeavors, although German involvement in international scientific enterprises did not necessarily depend upon expectations of material gain.[10] Indeed, prestige became an important factor in attempts to place Germany on equal footing with other major powers. After all, there was the example of the 1874 Austro-Hungarian expedition to the Arctic Circle, which had yielded both scientific results and the naming of an island after Emperor Franz-Josef.

Finally, polar ventures, like all expeditions to exotic places, attracted government support because of the popular interest in what was likely to become an epic adventure. The idea of polar exploration was no less fascinating to the turn-of-the-century public than to government officials and scientists. Readers were captivated by the Norwegian Fridtjof Nansen's account of his exploration of the Arctic Circle, translated into several languages (including German), while geographical games involving a race to the Pole became a popular form of entertainment.[11]

Public support for a German airship expedition to the Arctic grew steadily. Caricatures depicted the Zeppelin as claiming northern territories for Germany over the heads of other nations.[12] The imagination of air pioneers, intent on succeeding by any means available, reinforced the public's willingness to believe in the possibility of such endeavors, and vice versa. Several books pushed the idea of a lighter-than-air polar exploration, claiming that a successful flight up north would further prove the soundness of the airship formula.[13] Patriotic associations also became involved. The Zeppelin Bund, one of many groups formed to support airship development, popularized the idea of an exploration of the North Pole.[14]

In a related development, the beginning of the international tourist industry led certain fortunate travelers to Scandinavia. This early version of the package tour allowed travelers to follow in the footsteps of the romantic author Ernst Arnd. Scandinavia, along with Switzerland, became a preferred destination of the German bourgeoisie on holiday.[15] For the fortunate few who could actually journey to the departure base used by Arctic pioneers and heroes, the experience contributed to popularizing further the idea of a flight to the Arctic.

Last but not least, the optimism associated with flight technology in general predicted wondrous cultural feats, such as the spreading of Western enlightenment to other lands and a shrinking of the globe. All aspects of a *Kultur* society would benefit, from defense and commerce to science. "Peary's trip was purely a sports feat with no scientific value," claimed one observer, who also noted that scientists would benefit immensely from aeronautical polar research as it related to meteorology and geodetics.[16] Time and again, however, aviation supporters made sensational claims about how the airship would be "the polar explorer's means of transportation."[17] Such wild promises, at a time when merely flying in a rainstorm could prove deadly, nonetheless attracted attention and overshadowed calls for the expansion of "serious science"; adventure counted for more than patient endeavor. Thus, public enthusiasm for polar adventure by means of an airship was high. The airship and its chief advocate were waiting in the wings.

Count Zeppelin had been fascinated by the writings of Postmaster General Heinrich von Stephan, an early proponent of airmail who had argued the benefits of a polar route.[18] In 1907 the count drew up plans for an airship expedition to the Pole and enlisted several luminaries to advise him on the best way to reach it. The expedition would be financed in part through airship tourism trips to Scandinavia.[19] Two years later one of the count's associates, the meteorologist Hugo Hergesell, traveled to Norway to discuss the plan with Norwegian experts.[20] Noting the trip's political overtones, the German embassy in

Oslo remarked that Hergesell's visit had become an occasion for the local press to praise the achievements of German science in solving the problems of flight, thus boosting Germany's reputation as a leader among nations.[21]

In the spring of 1910, several developments accelerated the pace of the planning. Experts in northern German industrial circles studied the project carefully; it was agreed that Hamburg was the obvious starting base from which to reach the Arctic.[22] That summer, the count joined Hergesell on a trip to Spitzbergen aboard a government ship to study weather conditions there.[23] Upon their return, a book entitled *With Zeppelin to Spitzbergen* was published, its main thesis being that the airship could make invaluable contributions to the scientific exploration of the polar region (for example, by taking measurements for such purposes as hydrography). Emphasizing the "honor for German science," the book roundly criticized the project's opponents as adversaries of progress.[24] The introduction to the book was written by Prince Heinrich of Prussia—like his brother the emperor, an enthusiastic supporter of new technologies. The prince had accompanied Zeppelin and Hergesell on the trip to Spitzbergen. In his introduction he emphasized that the polar airship project would be serving the goals of science and fulfilling a "cultural duty . . . worthy of the German name."[25] National prestige was the uppermost concern, the flashy aspects of exploration being downplayed so as to emphasize the seriousness of the project.

Cautionary voices made themselves heard as well, Hergesell's being one of them. In a separate article that seemed to belie all his previous enthusiasm, the meteorologist pointed out that to reach the Pole would require a machine with a forty-eight-hour flight autonomy; no airship of the time was capable of such performance. Several other observers, relying on evidence from earlier failed attempts, echoed Hergesell's caution in popular scientific publications and warned that misguided enthusiasm could only lead to accidents.[26] Nevertheless, faith in the new technology prevailed, and the count had a permanent meteorological station established at Spitzbergen in 1912.[27] Two years were then spent collecting weather observations in the hope of mapping out the best possible conditions for an airship flight. In 1914 the count unveiled plans for a flight to Alsten Island off the Norwegian coast as a test for airship operations in cold weather.[28]

Unfortunately, World War I interrupted all work on the Arctic project. Air activities in the polar region, whether by plane or by dirigible, were not to resume until the mid-1920s, when plans developed immediately after the war to travel to both poles were revived.[29] Admiral Richard Byrd's successful 1926 flight to the North Pole by airplane did not dampen the enthusiasm; the Arctic had indeed been "vanquished" by air, but without in-depth scientific surveys, it remained a mystery. Nevertheless, the continued enthusiasm for Arctic exploration did not translate into success for the airship. Its technological prestige had declined in the face of World War I. The concept of a dirigible as an instrument of peace, progress, and science needed to be reestablished.

The year 1919 saw several efforts undertaken to keep the airship alive and increase its value as a civilian machine. While representatives of the Zeppelin company negotiated for sales in the United States or sought to diversify the company's industrial applications, several Zeppelin pilots explored other options. For example, retired captain Walther Bruns suggested linking the continents through a polar route.[30] Bruns's speeches aroused interest, but in fact he had little more to offer than ideas and enthusiasm. His letters to the Zeppelin company reminding it of its founder's ideas on the subject failed to attract attention.[31] Even the publication of Bruns's proposal in the periodical *Nachrichten für Luftfahrer,* which was sponsored by the Transportation Ministry, evoked only a limited response.[32] Most experts thought of airship links on an intra- rather than intercontinental level, despite the wartime distance records German airships had set.[33] The only bright spot for Bruns was his success in attracting the attention of several members of the Swedish Academy of Science, including the sinologist Sven Hedin and the famed polar explorer Fridtjof Nansen. Nansen had found the project somewhat fantastic, but contact with members of the German scientific community soon convinced him of the possibilities of scientifically investigating uncharted polar areas by air.[34] Nothing more, however, could happen until the airship had proven its worth in peacetime conditions.

The trend shifted dramatically with the October 1924 delivery of the ZR 3 airship to the United States as a war reparations payment. The successful Atlantic crossing unleashed a wave of excitement born of repressed pride and resurrected memories. In nonpolitical terms, the accomplishment suddenly gave substance to Captain Bruns's claims about the airship's long-distance prowess.

By the time ZR 3 landed in America a new scientific group, the International Society for the Exploration of the Arctic with Airships, or Aeroarctic, as it came to be known, had made public its purpose and goals.[35] The formation of such a society now, at a time when the political situation of the Weimar Republic had barely stabilized and government and other support would be difficult to obtain, was an audacious step. Aeroarctic was a consortium of mostly German scientists (fifty-eight of the eighty names listed), with Scandinavian scientists representing the second largest group (fifteen). The expulsion of Germans from learned groups as a result of the war explains why there were no French or British members in the society, and only one American. Among the Scandinavian presence, the only prominent link between Aeroarctic and the Royal Swedish Academy was the explorer Sven Hedin, who had previously expressed interest in the Bruns proposals. Hugo Hergesell, who had helped Count Zeppelin scout Spitzbergen before the Great War, was one of the society's founding members.

Aeroarctic represented a special mix of political nationalism and scientific internationalism. This quasi ideology involved exploiting scientists' sense of duty to their profession in such a manner that although information might be traded internationally, scholars would still work toward earning success for themselves and, in turn, for their country.[36] The politics of science could also affect diplomatic relations. The fact that the Swedish academy, a highly respected neutral organ, sent a congratulatory telegram to Eckener lauding the value of the airship caught the attention of the German Foreign Ministry.[37] It was becoming apparent that science might function as a kind of "power substitute" that could serve the German government's interests in a time when many avenues of political influence were closed.[38]

Hedin's sudden enthusiasm intrigued the German embassy in Stockholm, although the particular project that interested him—using an airship to study certain regions of Asia—conflicted with Bruns's plans for going to the North Pole.[39] As proposals for using the airship as a research tool, both plans had merit, but they were impractical and their realization unlikely, especially in terms of funding. Even so, reports that an Arctic exploration airship might be built in a foreign country alarmed the German ambassador in Stockholm, who asked the Zeppelin company for information about Bruns. The ambassador also noted that building the airship abroad would send two negative signals about Germany: first, it would hurt the industrial potential of the Zeppelin company; second, the Allies might think that Germany was implicitly accepting the terms of the Versailles treaty calling for the dismantling of airship installations on its territory.[40]

Informed of the matter upon his return from the United States in November 1924, Hugo Eckener discounted Bruns's initiative, claiming that he himself had suggested such an expedition to Roald Amundsen in New York but that negotiations had collapsed as a result of a press leak.[41] Eckener emphasized that the scientific aspect of the idea would help keep the airship in Germany, adding optimistically that "even France could not oppose the building of such a science-oriented airship without in turn taking a stand against *Kultur*."[42]

In the Aeroarctic project, then, matters had come full circle from the airship's pre–World War I origins. Pragmatic issues of commercial success, industrial potential, and power status had become intertwined in such a way as to reconvert the airship into an instrument of peace and national pride. This role also involved serving science as part of the *Kulturtat* along the lines Prince Heinrich had advocated.[43] However, politics were never out of the picture. With France intent on seeing the Zeppelin company's potential destroyed, it was a matter of honor for Germany to save from Versailles' claws what had been "until now a German specialty."[44] As a result, the German Foreign Ministry advised its representatives in Scandinavia to start a campaign to discredit any foreign construction plans, at least until matters could be clarified between Eckener, Nansen, and Hedin.[45] Intra-German preferences were also involved in the choice of a builder. Although it soon learned that Johann Schütte of the Schütte-Lanz airship company wished to compete for the potential contract, the government continually referred to Friedrichshafen rather than Mannheim (Schütte-Lanz's base) as a construction site.[46] Should the polar airship project succeed, it would be not just a German undertaking, but a Zeppelin one.

Conflicting Projects

The arrival on the scene of Eckener and his cordial meeting with Hedin further contributed to the impression that the Zeppelin company and its creations were important for the international scientific community.[47] All of Hedin's statements were scrupulously relayed to Berlin, from his longing to retrace Marco Polo's steps to the fascination the airship's passage would have for all "Oriental" peoples. The "giant dragon," as Hedin fantasized, would be the ideal instrument for a historic feat: charting the last of the globe's blank spots while restoring the German people's pride and self-confidence.[48] As for Aeroarctic's German section, it understood the possibility of using the airship "cultural route" to get rid of the "Versailles chains."[49] Here Aeroarctic echoed the physicist Max Planck's view that science (and in turn *Kultur*) was the one great achievement Germany could still lay claim to.[50] Shaking off the "barbarian" tag hung on Germany during World War I would serve the nation on all levels.

Eckener visited Sweden to address many public forums on the topic of airship flight. This visit inaugurated a new phase in the interplay of public support and official decision-making in the history of a project with commercial, political, and scientific considerations.[51] Until now, Bruns, enthusiastically supported by the Norwegian polar explorer Fridtjof Nansen, had been spending a great deal of time in the limelight publicizing the idea of airship travel to the North Pole. Bruns had proposed to bring in the Soviet Union as a direct partner in the venture. The Russians were at the time very interested in the use of airships for long-distance links across the Soviet Union, and they had previously collaborated with Germany on civilian airplane links as well as aeronautical construction. However, the German Foreign Ministry had worked since early 1925 to quash Bruns's idea about involving the Russians. Indeed, both the German public and the Ambassadors' Conference (the organ in charge of enforcing the provisions of the Versailles treaty) might misinterpret such a move.[52] In effect, then, any hopes for support required approval or even financing from the German federal government first.

Under article 142 of the Weimar constitution, the Reich and Prussian Interior ministries remained in charge of all funding for scientific projects, as they had in imperial times. The newly created Transportation Ministry was in charge of direct contacts with the aeronautical industry, including Zeppelin (since officially, at least, Germany no longer had an air force). Since the Arctic expedition was both a scientific and a transportation matter, both ministries became involved in setting the official conditions for support of Aeroarctic. In addition, Reich subsidies and funding were subject to ministerial screening and parliamentary approval. Finally, the Foreign Ministry continued to be a player, as the original idea had the enthusiastic support of the German envoy to Sweden, and the permission to build the airship very much depended on continuing negotiations at the Ambassadors' Conference in Paris.

With three ministries directly involved with the project, conflict was bound to arise. The first stumbling block concerned the initiators' failure to unite behind a single plan. In line with Hedin's proposals, German officials, who had earlier sought to kill any proposal for an internationally built airship, now told Nansen to internationalize the notion of Arctic exploration as much as possible, and to emphasize the purely scientific purpose of Aeroarctic.[53] The apparent contradiction inherent in an international undertaking having national goals dated back to the Versailles treaty restrictions: without at least the token involvement of an Allied power, the airship would disappear.

In the meantime, Eckener's attention was focused on the bottom line. All these debates over direction and priorities had done nothing to resolve the Zeppelin company's precarious financial and legal situation.[54] By the spring of 1925, Eckener felt that the German federal government was not helping the Zeppelin company the way it should; despite the ZR 3 success, the government was concentrating on airplanes.[55] Furthermore, the prospect of having to go through Aeroarctic to get any national or international attention was quite unpleasant. Eckener warned Transportation Ministry officials that he would sever all contact with Aeroarctic if they did not muzzle Bruns, who carried on about airship matters to anyone who would listen. Within days, the Transportation Ministry promised that matters would be worked out, asking in return that Eckener stay in touch with Aeroarctic at all costs.[56] Nansen, who visited Berlin early that summer, showed goodwill, although he did not understand why Bruns posed such a problem both to Berlin officials and to Eckener. (The fact that Nansen had delegated to Bruns the power to negotiate on his behalf complicated matters all the more.) Meanwhile, Sven Hedin, who was already considering other alternatives for an air expedition, had obtained Eckener's agreement that an airship trip to Asia would take place after the first polar exploration flight. That understanding was reached without consulting Bruns, further confusing the community of explorers over the various projects.[57]

Following his visit to Berlin, Nansen, in an attempt to comply with the Interior Ministry's requests, considered interceding with the French representatives to the League of Nations to obtain permission to build the airship.[58] Were the Allies to change their mind on the airship issues, then a basis would exist for seeking international funds.[59] The trick would be to convince Germans that a token degree of foreign scientific involvement was essential. The German public would not accept the notion of another "international" dirigible in the works, since that might represent acquiescence in the Versailles status quo. In that respect, national technological power took priority over any kind of scientific endeavor. On a more personal level, Eckener may have seen the move as a way finally to exclude Bruns from the proceedings. Nansen and other members of Aeroarctic noticed and deplored the turn of events toward Eckener's nationally centered plans, and sought to counter it and maintain the organization as an international effort.[60]

Matters came to a head in June 1925. Attempting to mediate between the two parties, Transportation Minister Krohne urged a Berlin meeting between Eckener and an Aeroarctic committee.[61] At the end of the stormy gathering, the Zeppelin company agreed that it would build the airship and that Bruns would be one of the crew members on the polar expedition.[62] Shortly after, Hugo Hergesell, who had resigned his membership in Aeroarctic earlier that year, publicly attacked Bruns as a plagiarizer of Count Zeppelin's ideas and Aeroarctic as a gathering of amateurs that used forty-year-old charts to study flight possibilities. Eckener seized the occasion to further question the seriousness and scientific basis of Aeroarctic's reports, but the press did not make an issue of this fact, emphasizing instead the tentative agreements Eckener had reached with Aeroarctic.[63]

In the meantime, the pace of polar successes was accelerating, which in turn intensified the sense of urgency surrounding the airship project. The Norwegian explorer Roald Amundsen flew to the Arctic Circle in a German Dornier hydroplane in July 1925.[64] He actually helped the Zeppelin project by commenting that in his view the airship would be better suited for such a task than his aircraft had been. He added that Eckener's project was the best proposal on the table at present, and other fliers, including Lincoln Ellsworth, reiterated this view, without even mentioning Aeroarctic. The German press eagerly picked up Amundsen's comments: that a personality unfriendly to Germany would speak sympathetically of a German hero seemed to confirm both the viability of the idea and the interest of the international airship community in a polar flight.

By the summer of 1925, then, the reclamation of the airship as a German achievement was well under way.[65] Eckener now moved to exploit the popular support for the "German" airship to structure the project according to his vision. He perceived that he was operating from a position of strength, despite the lack of financial support from government sources. This explains his move to initiate the Zeppelin-Eckener *Spende*, a popular fund-raising drive along the lines of the one that had saved Count Zeppelin's work back in 1908. Eckener's public standing was high, and popular pressure might finally succeed in forcing the reluctant authorities in Berlin to approve the construction of a new Zeppelin airship.[66] Early leaks in the German press about the *Spende* led to a discussion about a connection between the fund-raising and the polar exploration plans. Bruns, in the Soviet Union at the time, seemed to support the venture,[67] although behind the scenes both he and the German section of Aeroarctic were scrambling to preserve their influence in the polar airship project.[68]

After Eckener overcame the government's opposition to the fund-raising drive, he learned that Allied approval for construction of a German airship, even for scientific purposes, was by no means assured; articles in the French press had started accusing Germany of seeking to undermine treaty provisions.[69] It was essential, then, to avoid the nationalist frenzy that a public fund-raising effort might entail. The German embassy in Stockholm even suggested that the *Spende* be internationalized and pushed back to a later date. This ran counter to the need to maintain German public interest in the matter.[70] The publicity surrounding the Zeppelin jubilee celebrations was already competing with other news. Roald Amundsen's plans to fly to the North Pole in a semi-rigid Italian-built airship, the N 1 *Norge* (Norway), had attracted considerable public attention. The Norwegian-Italian venture to the Pole—General Umberto Nobile was to be the airship pilot—sought to prove that aviation could help solve scientific problems and promote other successes.[71]

The Versailles treaty restrictions, however, continued to hamper German efforts. Criticism of the Allied position emphasized the misguidedness of Allied policies and the need for a more global perspective. Delaying the accomplishment of the *Kulturtat* would not only hurt Germany but would hold all of Europe back in the field of commercial aviation. Several luminaries attending an aviation conference in Dessau in September 1925 praised the flying machine as another tool of international transport.[72] They welcomed with particular enthusiasm the anticipated provisions of the Locarno treaty and predicted that the new atmosphere of relaxed tensions would enable Germany to contribute to aviation and thus reestablish its reputation for furthering the causes of science and civilization.

The Zeppelin-Eckener *Spende* proceeded from late 1925 until early 1927, during which time Eckener and several airship captains gave lectures throughout Germany intended to raise public awareness of the project. In order to maintain the goodwill inspired by the initial rumors of an Arctic flight, Eckener acknowledged the possibilities of such exploration.

In the meantime, Aeroarctic went about gathering support for its own plans.[73] By the fall of 1926 the society had resolved some of the problems the Eckener-Bruns strife had caused. Members very much counted on the name of Fridtjof Nansen, the society's president (who had earned the Nobel Peace Prize), to lend credibility to the project. There was great hope that his reputation would facilitate any discussions with the German government about building the airship.[74] However, Aeroarctic failed to get full support from all the scientific circles it contacted; more important, there was no single office in Germany on which to concentrate its lobbying efforts. The structure of

Zeppelin's hypothetical arrival at the Pole. While the count did visit Spitzbergen off the Arctic Circle, he never saw the completion of an airship expedition to the Pole. This postcard satirizes the craze surrounding the count's persona, suggesting that even indigenous fauna would initiate a celebration upon hearing, "Zeppelin is coming!" ASTRA

Aeroarctic and the nature of its goals required dealing with at least three different Reich ministries as well as with the Prussian government. Aeroarctic did have the inside track at the Interior Ministry, whose leader, the left-wing Liberal Wilhelm Külz (who before World War I had flown aboard a Zeppelin), was quite eager to help Nansen.[75] Külz actively supported Aeroarctic's activities throughout his two terms as Interior Minister.

This inside track caused a clash between the Transportation and Interior ministries in early 1926 over issues of responsibility. Transportation Minister Krohne stressed that "only once these are resolved do the scientific duties of the scholarly society [and therefore of the Interior Ministry] come to the foreground."[76] The matter was something of a tempest in a teacup, but it also involved two individual politicians vying for the attention of an international luminary, Nansen. Krohne personally promised Nansen that the German government would help build the airship and make it available for two flights in 1927.[77] In any case, the Transportation Ministry successfully reasserted its power. In effect, the scientific implications of Aeroarctic's airship proposal once again took second place to economic and political considerations.

The Foreign Ministry, because of its responsibility for defending German air interests wherever the Versailles restrictions were enforced, paid close attention to Aeroarctic. Foreign Minister Gustav Stresemann, despite cold-shouldering Eckener, had hailed the flight of ZR 3 to the United States in October 1924 and held it up at the Ambassadors' Conference the following year as an argument for a loosening of the Versailles restrictions on aviation.[78] He did not succeed. Ironically, by the time the Ambassadors' Conference finally decided to specifically lift restrictions on airship building in the spring of 1926, the Foreign Ministry had to admit that it had lost interest in the polar undertaking.[79] In Friedrichshafen, Eckener, who had accused Aeroarctic of having done little to help relax the Versailles restrictions, also seized the opportunity to distance himself from both that group and the government.[80] When Aeroarctic held its first general assembly in the fall of 1926, Eckener's renewed opposition to its plans diminished its apparent success in obtaining governmental support.

Inaugurating the Aeroarctic meeting on 10 November 1926, Fridtjof Nansen attempted to drum up both popular and scientific enthusiasm. "We accept the word *fantastic*," said the Nobel Prize winner. "Only research inspired by imagination is truly productive; only it can courageously move forward."[81] Unlike Prince Heinrich's words some fifteen years earlier, which had downplayed the excitement of exploration, Nansen's speech reminded the thirty-five representatives of the foreign and national press that enthusiasm did have its place in human progress, especially in a case like Aeroarctic's, in which success was tied to a very limited budget. The state of the German government's finances meant that scientific grants were limited, and funds funneled through national associations, such as the Kaiser Wilhelm Society, went to individuals rather than research centers.[82] The Emergency Association of German Science, supported by industrial funds, was more likely to offer assistance to national projects such as the German Arctic expedition than to an international body.[83] Under these circumstances, Aeroarctic pinned its hopes on public and even international support.

To ensure ongoing interest, Aeroarctic needed to clarify how and for what purpose it intended to use the airship. Between the publication of its organizational pamphlet in the fall of 1924 and its first general assembly two years later, Aeroarctic had evolved from a group of mostly German and Scandinavian scientists into a truly international body that included members from the former enemy nations. By 1926 its membership included airship and airplane pilots from several countries; General Lucien Delcambre, the French aeronautics inspector, served as one of its four vice presidents.[84] Such a glittering membership was deemed necessary because Hugo Eckener had distanced himself from the group's activities. Aeroarctic dared not return the rebuff but even had to acknowledge Eckener's achievements at its conference, since the new Zeppelin airship was to serve for two flights to the Arctic. In effect, the Aeroarctic board was deliberately overlooking Eckener's sulky behavior, forcing him into the position of benefactor. This was not an altruistic move, since negotiations with Transportation Minister Krohne on the construction of an airship for research purposes had led the minister to insist on involving the Zeppelin company, and thus Eckener.[85]

As a media thrust, Aeroarctic's 1926 meeting was not a resounding triumph. Several sessions witnessed members fighting publicly over the relative merits of the airship versus the airplane as a research tool, in the wake of Richard Byrd's recent flight to the Pole; and Sven Hedin, previously a staunch partisan of the airship, had already turned to the newly formed Lufthansa airline for help in his exploration of the Far East.[86] The debate was a continuation of the traditional competition between airplane and airship pilots, each side insisting on the superiority of its particular kind of machine. Partisans of the airship apparently carried the day, however, relying on the fact that Amundsen and Nobile had plans to fly to the Arctic in an airship, and that the British air minister had just given preference to the airship over the airplane for covering long distances. Furthermore, the vaunted role of the dirigible as an instrument of "international understanding and peace"[87] gave airship supporters the moral advantage, although this claim turned out to be a double-edged sword.

First, the traditional competition among airship types resurfaced. Claims by the airship builders Johann Schütte and August von Parseval that rigid airships (of the Schütte-Lanz wooden type) and blimps both had as much potential as the Zeppelin for Arctic exploration further weakened the position of the Zeppelin supporters in Aeroarctic. Schütte and Parseval both clearly hoped to revive their defunct factories, but Aeroarctic failed to use this as a weapon to tame an overly confident Eckener. Instead, this temporary challenge offered further evidence that Aeroarctic was having trouble defining its priorities.

The image of Aeroarctic as an international body using the Zeppelin as an instrument of global understanding disconcerted those who saw the airship as a powerful symbol of German revival. The flying machine, in whatever shape, remained a national flag carrier, even when crossing geographical frontiers in apparent freedom.[88] From this perspective, Aeroarctic's international framework was a liability rather than an asset.

By the time a second meeting of Aeroarctic took place in September 1927, Eckener's position had weakened. His fund-raising drive had fallen short of its goals, leaving him in an embarrassing position: despite an insufficiency of funds, construction of the new Zeppelin was already under way. The Reich had initially contributed 400,000 marks, but it was dragging its feet about granting additional help. One possible way out of the impasse required Nansen's help. The president of Aeroarctic agreed to seek further international support. A review in the spring of 1928 to consider the fruits of his efforts would likely clear the way for the two previously guaranteed polar flights in 1929.[89]

In the meantime, public interest remained high. Amundsen had succeeded in steering the semirigid *Norge* to the Pole. General Nobile, one of the leaders of the expedition, returned home in triumph. On the prompting of Mussolini and the Italian air minister, Italo Balbo, he then prepared to initiate a similar flight on board the *Italia* dirigible, a sister ship of the *Norge*. The *Italia* left Rome in May 1928 and flew northward to Spitzbergen Island. On 25 May, after circling the Pole, the machine crashed. The control car was torn from the hull and flung off, carrying several crew. Those in the control car, including Nobile, survived and were found a month later. The loss of the Italian airship, blamed on human error (Nobile resigned his commission in disgrace and wasn't rehabilitated until 1978), cast a shadow over the reputation of the dirigible as a means of polar exploration.[90]

Consequently, Aeroarctic's future plans remained unclear because of limited international funding. Although the society had succeeded in getting some research funds from the Carnegie Institution, it could barely cover its operating expenses. The German government provided additional support by issuing public statements in the summer of 1928 suggesting that the two Arctic flights would still take place at no cost to Aeroarctic. That decision represented a curious compromise between the competing interests of the German Interior and Transportation ministries. The Reich government agreed to help Zeppelin complete the new airship (fulfilling the Transportation Ministry's wishes) in return for the promise that the ship would undertake the planned Arctic flights (as desired by the Interior Ministry). Yet such deal-making left Aeroarctic with more financial problems than it solved.[91] True, the government had promised to help Aeroarctic, but it had also imposed major conditions. While the flights themselves would cost nothing, the society would be responsible for logistics, such as the acquisition of scientific measurement tools and proper insurance. Furthermore, it would have to come up with the funding for these preliminary costs by January 1929.[92] In effect, although perhaps not intentionally, the German government had just ensured the failure of Aeroarctic.

In a development that seemed to embody the proverbial "for want of a nail" syndrome, Aeroarctic foundered on the issue of commercially insuring its endeavors. While funding might be found to pay for the flight itself and scientists were available on a voluntary basis, there proved to be no way for the society to raise the money to cover the insurance premiums. Already high to begin with, premiums had shot up following the *Italia* crash.[93] In a further catch, Aeroarctic was forbidden to issue public appeals in a foreign country (such as the United States) and could accept only unsolicited offers of foreign aid. This also meant that the society could not ask for support from foreign sections of Aeroarctic, for fear that this might damage Germany's precarious sovereignty.[94] As a result, while Aeroarctic asked for a postponement of the deadline for obtaining the necessary funds, Eckener busied himself in the fall of 1928 with testing the new *Graf Zeppelin* airship. The success of the venture gave both Eckener and, paradoxically, Aeroarctic new maneuvering space.

Selling Science as Entertainment

The publicity surrounding the *Graf Zeppelin*'s first flight across the Atlantic provided a show of official and public goodwill for Eckener and Aeroarctic. While Aeroarctic considered new options such as a joint venture with the Soviet Union using aircraft instead of a dirigible, the German government now promised to cover all costs for at least one voyage to the North Pole, in fulfillment of its agreements with the society.[95] On paper, however, plans favored a *Graf Zeppelin* expedition and finding nongovernmental funds to cover ancillary costs.[96] Captain Bruns reappeared on the scene as a fund-raiser for Aeroarctic. As was often the case in the story of the Arctic expedition, Bruns mixed official and media activities. He and Nansen traveled together to the United States in January 1929 to negotiate for the use of an Alaskan base. Bruns used the occasion to sign a contract with the Hearst news organization that called for a meeting of the airship with a submarine in the Arctic Ocean to exchange mailbags. In addition to income from this publicity stunt, other funds would come from the German popular press, and from advances for Nansen's book, to appear soon after the first expedition.[97]

Newspapers kept announcing various flight plans to the Arctic and other destinations, each story relying on different sources. The Soviet Union was acknowledged as an intermediate base for any Arctic flight, but details varied for lack of concrete answers from Bruns, who was actively negotiating with Soviet scientists and authorities.[98] By this time, however, the German government, interested in building goodwill abroad, where the airship was regarded as a technological marvel, wanted a visit to a populated area rather than a flight to an ice desert. Eckener agreed readily, for business reasons.

Bruns remained busy assembling the necessary funds to pay for the Arctic flight and insurance. Only the Prussian Ministry of Education offered a token contribution to keep Aeroarctic afloat.[99] Had Nansen been in charge of all the negotiations, things might have been different, but Bruns was an erratic representative for Aeroarctic. In yet another ill-conceived move, he canceled negotiations with the Ullstein press group and the WTB wire agency, enterprises favorable to the Weimar Republic, and almost concluded an exclusive deal with the Scherl publishing group, a property of the right-wing Hugenberg concern. The Reich government, fearing both domestic and international repercussions, stepped in and demanded that Bruns reopen negotiations with the liberal publishers.[100] Meanwhile, the unresolved insurance issue became public knowledge, further complicating Bruns's and Nansen's endeavors.[101] Finally, the stock market crash further increased the challenge of finding both sponsors and willing insurance companies.[102]

In desperation, Aeroarctic asked the German government to vouch for the society's integrity.[103] Trying to refocus the issues at stake, Külz, the former interior minister who had actively supported the group's efforts and who was now a member of the Parliament, called on his successor to help save the enterprise. As Külz pointed out, even though the society was an international one, its demise could have negative effects on Germany, especially if Nansen were to come out of the affair with his reputation tarnished. In short, Germany had to underwrite such a *Kulturtat*.[104]

In December 1929, following discussions between Nansen, representing Aeroarctic, Sir Hubert Wilkins, who had flown to the South Pole, and Eckener, another meeting between Aeroarctic and government ministries took place in Berlin. By now Eckener had decided, following his round-the-world flight the preceding summer, that a polar exploration was too dangerous.[105] He nevertheless agreed that one flight remained a possibility and that the Zeppelin company rather than Aeroarctic would attempt to secure an insurance policy to cover it. All Aeroarctic could bargain for was yet another rescheduling of the flight for sometime in 1931.[106]

Nineteen thirty was a year of heightened activity for the *Graf Zeppelin* and for Arctic exploration in general. Despite the deepening economic turmoil, the airship flew throughout Europe and back to America. Alfred Wegener, who had suggested the continental-drift theory back in 1912, led a new German land and sea expedition to Greenland, and Professor R. L. Samoilowitsch, an active member of the Russian branch of Aeroarctic, drew up plans for an expedition aboard the icebreaker *Krassin*.[107] There was silence on the matter of the airship trip to the North Pole. One scientist publicly accused Eckener of manipulating the polar exploration issue in order to get his airship built, but the incident had few public repercussions.[108]

In May 1930 Aeroarctic suffered a further, and definitive, setback. Exhausted by his international work and reportedly upset at the ongoing difficulties with funding,[109] Fridtjof Nansen had a heart attack and died. The society lost its driving force and most prominent spokesman. No other member of the society had the prestige required to steer an international group, especially in such difficult economic times. Under the circumstances, Bruns, as secretary general, felt that he had to turn to his old nemesis, and he offered the presidency of Aeroarctic to Eckener. Eckener accepted on the condition that the society's general assembly confirm his appointment, which it did in October 1930.[110] For Eckener, a former threat had been transformed into a means for furthering his goal of promoting the Zeppelin. Yet it was a sputtering vehicle at best; by December, Aeroarctic lay in financial ruin, unable even to support the publication of its journal.[111] Eckener could do as he pleased, but the Zeppelin company was also saddled with costs that Aeroarctic could not assume.

Eckener, as always, used his skills at managing the media and his reputation as a shrewd businessman to remedy the situation. A well-planned press campaign whipped up public enthusiasm for the polar expedition. "After Twenty Years, Count Zeppelin's Plans to Become Reality" ran a headline in July 1931.[112] Eckener allowed three weeks to prepare the *Graf Zeppelin* for its polar trip, scheduling flights to Iceland and Norway to test weather conditions in colder regions. Both flights had obvious value as training exercises for an Arctic flight, although they were presented as courtesy visits to northern nations.[113] The German Aviation Insurance Consortium resolved the insurance matter by agreeing to accept payment after the polar flight. This would allow Zeppelin to collect the moneys from the press and mail contracts before settling the 2-million-mark bill.[114] Eckener also succeeded in signing sponsorship deals with German companies. Enterprises as varied as I.G. Farben, Zeiss, and Knorr supplied the expedition with scientific materials and other goods. In contrast, involvement with official circles was kept at a minimum.

During a cabinet meeting in July 1931, the Reich ministers voiced concerns about the physical dangers and political consequences of such a flight. If it failed, the expedition would damage Germany's image and the politicians would suffer. Yet the odds were against the government's being able to stand in the way of the expedition. Eckener was in charge of the operation, technical and business interests had invested in the enterprise, and the scientific side remained international, with several Russian scientists leading the learned group that was to fly on the *Graf Zeppelin*. Were the airship to crash, would it also irreparably damage the image of the Zeppelin company? The Finance Ministry in particular was worried about answering why it had subsidized the company. However, it became clear that since most of the financial support for the expedition was coming from private and business sources, Berlin could not use financial leverage to force the cancellation of the flight.[115]

Eckener's major concern was media relations. He never permitted internal disagreements to filter out to the public (at least not from his side), preferring to project an easygoing yet serious image of himself and his activities. In one interview he explained that the flight to the Pole was simply the result of an accepted invitation.[116] In another the reporter described Eckener suspiciously examining an Aeroarctic map in his office, commenting that he was never one to trust a *Körperschaft* (corporation), whatever its purpose and reputation.[117] The image of the lone fighter reinforced the public's view of Eckener as a man constantly battling bureaucracies to fulfill the dreams of that other popular hero, Count Zeppelin. Justifying the mounting of the expedition during a time of such great national difficulties, Eckener claimed that the project had an international character. As if to echo his claim, the German media covering the expedition saw to it that the reporters included a Swiss and a Palestinian.[118] It also parroted Eckener's explanation that Germany considered the expedition to be not a luxury but a necessity to maintain its cultural presence. Last but not least, the money was already secured—good news for the taxpayer—and the country could not afford to lose face by backing out now with the *Graf* ready to go.[119] The patriotic note may have contradicted the emphasis on scientific internationalism, but Eckener was not one to worry about consistency. His primary goal remained the exploitation and development of the airship, not the furthering of science.

On 24 July 1931 the *Graf Zeppelin* left Friedrichshafen for Berlin, where several members of the Soviet scientific team boarded it.[120] Ullstein's correspondent on board, Arthur Koestler (who would later write *Darkness at Noon*), supplied the press pool with reports. The flight over Germany left in its wake "a wave of howling factory whistles, traffic snarls with madly hooting cars."[121] As Koestler pointed out, the media avoided mentioning that the airship was not going to the North Pole at all but rather to a zone prescribed by the insurance terms (when asked afterward about this matter, Eckener never answered).[122] The airship flew over Russia to Leningrad (St. Petersburg) and on to Nova Semlya, near Franz-Josef's Land. Radio reports stressed time and again how the scientific "pioneers" would make use of the airship's unique ability to succeed in their endeavors.[123] Not even the bleak economic situation at home, with banks failing and Germany trying to negotiate special loans from France and Britain, could divert attention away from the airship's voyage. In fact, reports on the expedition provided some measure of welcome relief. The degree of escape afforded by such reporting may have been limited, but the role the airship played, in this and other flights, as a source of universal entertainment should not be discounted.

The day the *Graf Zeppelin* landed on ice, the achievement shared the front page of the *Vossische Zeitung* with British prime minister Ramsay MacDonald's visit to Berlin. Mail was exchanged with the Soviet icebreaker *Malygin*—an earlier plan to rendezvous with the American submarine *Nautilus* was scrapped because of the latter's mechanical troubles—but the airship lifted off soon to avoid colliding with an iceberg.[124] The experiments conducted on board proved useful, although the details were ignored by the public.

Considerable photography and mapping took place during the Arctic flight. The airship's projected role as a measurement platform had become a reality. Measurements of the Earth's magnetic field were taken by means of experimental instruments on loan from the Carnegie Institution. The goal was to create navigational maps that would contain magnetic meridians, so as to facilitate compass navigation. The airship's quiet stability—so unlike the cramped and vibrating conditions on board an airplane—allowed precise measurements and mapping through a Lambert cylindrical projection.[125] The scientific team included eight Germans (only five of whom were actually scientists), two Americans (including Lincoln Ellsworth), three Russians, and one Swede.[126] Although the team retained the international flavor Aeroarctic had hoped for, it was smaller than originally planned. The heavy equipment and the requirement for a full crew complement limited the number of scientists who could be taken aboard.

These issues as well as the details of the experiments were entirely absent from public discussions about the flight. The intricacies of the experiments were too complicated to explain at length in newspaper columns. The only link to the scientists that the public could truly follow without journalistic assistance was the *Graf Zeppelin* itself, which of course could not reveal much about the skills of those on board. The airship expedition offered a new example of a phenomenon that had begun in the late nineteenth century: people accepted science but did not understand it. But whereas many technologies presented a complex and intimidating face to the outsider (consider the impression conveyed by a scientific laboratory), the airship was simple in appearance and pleasing to the eye.[127] Faced with this form of the technological sublime, the public was free to speculate on adventures and discoveries, and to imagine at will the experiments carried out in the giant's belly.

The return of the *Graf Zeppelin* to Berlin was also covered as an adventure. Some thirty hours after it had left the Arctic, an escort of several Lufthansa machines joined the airship as it flew into German airspace. The *Graf* then dipped over Berlin's Tempelhof Airfield and landed. Exhilaration swept the bystanders. Once again, technology had rendered humanity a great service by binding nations together and helping push back the limits of scientific knowledge. Celebrities like Admiral Byrd were on hand to greet the returning heroes, the great contributors to scientific knowledge, but it was Eckener, not the team of scientists, whom they hailed. Arthur Koestler set the tone of public appreciation: "We have double grounds to rejoice over this achievement, for it happened at a time when Germany is poor in material goods; but this poverty could not prevent Germany from giving the world the best of its own immeasurable richness."[128] Technological achievement thus came to represent a measure of the German soul. It was the product of hard work, relentlessly carried forward despite difficult conditions. The overarching goal of contributing to the international community softened the nationalist overtones of the achievement. If any obstacles remained, they were proving to be political, not technological. Eckener was quoted as being pessimistic about the future of transpolar flights, not so much because of weather or navigational problems but because of geopolitical issues involving "bordering nations."[129]

Following the successful completion of the Arctic flight and the triumphant reception of its participants in July 1931, the time was ripe for the planning of a second Arctic flight. But Aeroarctic could not provide any funding, and Eckener saw little advantage in becoming further involved in a venture that delayed the effective commercialization of the airship and jeopardized its safety.[130] He even asked the Transportation Ministry to quell rumors of another polar flight.

As the second International Polar Year dawned,[131] Germany's airship withdrew from any public involvement in the scientific investigations of the polar regions. With the economic crisis at its peak, additional flights would have entailed serious financial problems. In principle, the Soviet Union was willing to sponsor and even purchase an airship for the charting of the northern territories. Yet it too apparently lacked the financial means, as it still owed money for its participation in the July 1931 expedition.[132]

Moscow eventually made arrangements to settle the debt with the Zeppelin company, but with the Nazi seizure of power, talks of further cooperation in airship flights and of airship purchases came to an end.[133] As for Captain Bruns, he made one last attempt to exploit the propaganda value of a dirigible flight to the polar circle in February 1934. When Bruns learned that the Soviet icebreaker *Tscheljuskin* was in trouble near Wrangel Island, he wrote to Propaganda Minister Joseph Goebbels to suggest an airship rescue. Downplaying the costs of the operation and emphasizing the value it would have for Soviet-German relations, Bruns offered to lead the operation.[134] The rescue possibility was also discussed in the media, and there was some official interest in the matter, but no action was undertaken.[135] Once again Eckener had stood in Bruns's way, stating in a newspaper interview that an airship rescue mission would have been technically impossible: the *Graf Zeppelin* lay in a winter shed emptied of its hydrogen and undergoing inspection.[136] These reasonable technical objections allowed the German government to decline involvement in the affair (leaving Soviet fliers to conduct a successful rescue).

The episode was the last to involve any talk of Arctic airship adventures. Despite rumors of research expeditions published in newspapers,[137] no trip to the North Pole was ever again seriously considered. The *Graf Zeppelin*'s expedition to the Arctic Circle remained a single enterprise, bound to become the stuff of legends or to spark the imagination of writers. In December 1935 the German embassy in Brussels forwarded a request from the Belgian journalist Lucien Eichhorn to both the Air and Propaganda ministries; Eichhorn had produced a screenplay in which the *Graf Zeppelin* was to rescue members of a stranded polar expedition, and he was seeking the new Reich's support for production costs. Aware of the similarities with the Wrangel Island incident, the German government worried that the movie would actually glorify Soviet efforts at Germany's expense and declined assistance.[138]

In a contemporary account written for the general public, the anthropologist Ludwig Kohl-Larsen, a member of the *Graf Zeppelin* expedition to the Arctic, gave Hugo Eckener most of the credit for the expedition's success.[139] Kohl-Larsen also lauded public participation in the form of donations and stressed the selfless virtues displayed by the German people in hard times. In so doing, he ignored the other dimensions of the project, thus implying that they mattered little in the bigger picture. In fact, an odd mix of the politics of science, culture, and personality conflicts went into the success of the polar expedition, with various actors seeking to exploit the airship symbol and hoping for popular support.

The public had been fascinated with the idea of a polar airship flight for years before the 1931 expedition. From the very beginning the airship had been considered an achievement of science, and thus a palpable demonstration of what research could accomplish, but this attempt to put it to use as an instrument of science was only marginally successful.[140] Indeed, scientific knowledge was only an incidental by-product of the polar expedition. By Eckener's own admission afterward, the voyage had served primarily as an outlet for the popular ideology of adventure.[141] Perhaps the costs of mounting a serious scientific expedition would have been prohibitive, or perhaps there was simply insufficient faith in the value of the Zeppelin for scientific investigation. Aeroarctic's lack of clear goals and the political and financial obstacles it encountered were part of the problem as well.

Nevertheless, the scientific benefits of a Zeppelin expedition were stressed time and again before the flight took place, as if to prove how much science mattered. The benefits were emphasized in much the same way as the "science dividend" of the Mercury, Gemini, and Apollo space missions would be emphasized some three decades later, said dividend being dependent on public interest first and political interest second.[142] One might add that in Aeroarctic's case, Eckener's interests came first. The polar expedition may have yielded some new information, but it was primarily a triumph of technology for its own sake and for the sake of publicity.

The pattern paralleled that of scientific ballooning in America in the 1930s. There too the notion of a human presence in a forbidding arena was used to sell the idea of bona fide scientific investigation. Aeroarctic scientists happily joined the polar project, but found their interests subordinated to the business concerns of the society first and Eckener next. Their American colleagues faced a similar problem, whereby scientific results were valued far less than the human accomplishment.[143]

Scientific Zeppelinism provided welcome entertainment in difficult times, but it remained for the most part a resurrection of elements from the past. The long-forgotten dream of a dead hero was recast as a public memory and packaged as a kind of duty Germans had to honor, if only for the sake of nostalgia.[144] Though promoted as a scientific undertaking, the polar airship project provided little more than data on the machine's performance (it was found that the airship withstood very cold weather) and, of course, entertainment. Technology might not have been able to solve the nation's problems, but it could help people bear those problems by providing a glimmer of hope and a renewed sense of national pride. However, technology thereby displaced the goals of science and became an end in itself. The polar expedition may have contributed to the symbolic value of the airship, but it did not turn it into a useful scientific tool.

CHAPTER 6

Political Zeppelinism
Manipulating Airship
Culture, 1933–1939

N HIS MEMOIRS, HUGO ECKENER REMARKED ON THE
irony that the Nazi regime, which had made "Zeppelin thought"
triumph through expansion, in effect yielding to public opinion, also destroyed
both the business and the symbol.[1] Most histories of the rigid airship gloss over
such subtleties, however, detailing instead a few propaganda flights before
jumping to detailed analyses of the *Hindenburg* tragedy. They argue that the
airship fit perfectly into the Nazi cultural scheme to manipulate the masses: the
Nazis coopted the airship symbol, just as they had done with the Prussian eagle.
The airship's gigantism, its role as a redeemer of German honor, and of course
the military background of its inventor made it a perfect addition to the Nazi
pantheon of power signs until, in 1937, the destruction of the *Hindenburg* put
an abrupt end to the dirigible era.

Although factually correct, such analyses fail to note the ambivalent feelings
Nazis had toward the airship. The staggering popularity of the machine fed in
part on the old cult of Count Zeppelin; yet no German-born personality cult

could be permitted to share the scene with Adolf Hitler, let alone exceed his stature. At the same time, the Nazi regime poured millions of marks into the building of the *Hindenburg* and after its demise even initiated frantic searches for helium, hoping to save its sister ships, or at least their symbolic value. Indeed, the regime's emphasis on technology in propaganda literature suggests the importance of the machine's power as a symbol.

The Nazi airship riddle breaks down into three parts: the airship's role as a propaganda tool and related manipulations of the memory of Count Zeppelin; the machine's business and technological dimensions under Nazi conditions; and the management of damaged symbolism in the wake of the *Hindenburg* accident.

Nazifying the Airship

To understand the Nazis' difficulty with the airship, we must consider how totalitarian regimes deal with the symbols that predate them. A symbol always shrinks its message into its representation.[2] Because of this reduction, symbols come to carry unintended meanings.[3] Consequently, when regimes create a "symbolocracy" of national character for political purposes, they risk turning symbols into stereotypes.[4] Furthermore, once formed, a symbol may acquire its own dynamic, moving away from its original purpose toward an image that exists only for its own sake.[5] When this happens, the symbol does not automatically lose its earlier meaning; it may come to embody a set of contradictory values (the fact that the airship appealed to pacifists as well as militarists during the imperial era is a case in point). Thus, in Nazi Germany, where the airship figured prominently among the symbols associated with the Third Reich, the issue of emblematic value rises anew.

During the Weimar Republic, the airship blended successfully into the overall ambiance of modernity and unity. How would this trend carry over into the Nazi era? Whereas modernist culture was essentially elitist, Fascist views advocated a populist outlook. This raises the wider question of whether the Nazi regime was a modernizing force or simply the extension of a civilization in crisis.[6] Opponents of the modernizing view argue that modernization requires prior democratization. It follows that any concrete modernizing effects the Nazi era may have had should be seen as accidental rather than intentional.[7] Supporters of the modernizing view, on the other hand, claim that a free sociopolitical setting is not a prerequisite to progress, technical or otherwise.[8]

Whatever the merits of such a debate, the growing homogenization of German society under the Nazis actually removed the bonds of solidarity that had previously existed in Germany, and this was true as well in the case of the Zeppelin.

In the years of the Weimar Republic, Hugo Eckener had convinced many political personalities of the airship's potential and value, although the economic crisis prevented them from offering any practical support.[9] Now, under the Nazis, the Zeppelin firm was apparently winning concessions from the regime but was sinking under its control, not unlike other industrial concerns.[10]

On 19 January 1933, representatives from the Zeppelin company met with officials of the Labor Ministry to ask for an advance of 2.3 million marks.[11] The request had little to back it up; the requested funds would only keep a limited number of workers busy until both airship service and the world economy took off again. The German government had paid out some 650,000 marks in 1930 to keep the *Graf Zeppelin* flying and to honor foreign agreements such as the establishment and maintenance of installations in Spain and Brazil.[12] Other funds came from sponsored city visits, postal contracts, and the slow selling off of Zeppelin assets. For example, in 1932 Eckener ceded to Claude Dornier his aircraft factory (Dornier had started operations as a Zeppelin-owned subsidiary), and he approved plans for the city of Berlin to buy back the Staaken airfield where Zeppelin had assembled airplanes during World War I.[13]

As in earlier Weimar days, Eckener framed his request to the Labor Ministry in terms of the long-range benefits that would accrue to German business from such a transportation system. He also promised completion of the new LZ 129 airship for November 1933—an unrealistic claim in view of the slow pace of assembly.

By the time all the concerned bureaucracies had started reviewing the request, Adolf Hitler was the new German chancellor. The Zeppelin matter landed on the table of the Nazi aviation commissioner, Hermann Göring, who promptly rejected it.[14] Eckener later admitted that having finally won support from the previous regime, though too late to obtain funding, he had expected enthusiasm—and funding—from the new leaders.[15]

One month later, on 9 March 1933, reversing its previous attitude, the Air Commissioner's Office sent a letter approving funding on economic and political grounds. In view of British influence and American capital investments in Latin America, the psychological impact of regular German airship service to this region would be of great significance. The "highways of the sky" would also provide Germany with new avenues of foreign trade.[16] An official announce-

The view from above: the *Graf Zeppelin* visits Berlin during a rally, ca. 1934. Hugo Eckener, a conservative who admired Germany's imperial past, rejected Nazi policies and their related symbols. He therefore had his crew fly the old imperial German colors. Here, two mechanics wave the black-white-red flag from one of the engine gondolas, far above the giant swastika set up on the street below. *Illustrierter Beobachter*/RLM, 3236, via ASTRA

ment in early summer stated that the Zeppelin company would receive 2 million marks.[17] Propaganda Minister Joseph Goebbels remarked in his diary that this funding decision had about it a "feeling of sour grapes."[18] The remark neatly summed up the Nazis' ambivalent attitude toward the airship.

The political instrumentalization of the Zeppelin had been bound to occur, for like all aspects of aviation, the Zeppelin fit in with the new regime's desire to demonstrate its interest in technological innovation.[19] In fact, throughout the pre-1933 years, Nazi media had lauded the airship's success, using it as a prop in the Nazi myth of national rebirth. A picture of the *Graf Zeppelin*'s flyby during a trip to Munich was used to point to the "heroic" last stand of the 1923 Nazi putschists. The airship also offered a pretense for attacking Weimar democracy; its 1928 Atlantic crossing against heavy winds was likened to the "fight of the German people toward light." As one Nazi columnist noted, "The great publicity the airship is generating in the name of Germany is unfortunately pointless, so long as the likes of Stresemann . . . control German foreign policy."[20] Logically, then, the Nazis should have welcomed the opportunity to use the airship, but Hugo Eckener was a well-known opponent of theirs.

Eckener, who had intended to run against Hitler in the presidential race of 1932, had thus incurred the wrath of Nazi leaders even before they came to power. This was not the only black mark against him in the eyes of the Nazis. He was an opponent of the economic policy of autarky and of the regime's Jewish policies (which he viewed as economically counterproductive). Eckener was also incensed by the new flag order calling for a swastika to be painted on one side of the *Graf Zeppelin*, which he feared would be a source of bad publicity abroad. A traditional conservative, he preferred the red-white-black imperial flag. Not only did he encourage his crews to wave the imperial rather than the Nazi flag, but on the occasion of the airship's visit to the Chicago Exposition in 1933, he deliberately steered the machine in a semicircle that exposed only the side that bore the imperial colors.[21] He reportedly shifted his position after a meeting with Gestapo chief Rudolf Diels, who pressured him into obedience.[22] Diels could not, however, force Eckener to change his views. As Eckener wrote to his friend Arnold Brecht in exile in America, in 1934, "I know how great and widespread opposition to the new German political outlook is."[23] In short, as one Nazi regional leader commented, Eckener was "the least useful propagandist the new Germany could send to the world."[24]

Nevertheless, personal pique was one thing; the use of the airship for propaganda purposes was quite another. The funding gave the Nazi government access to the airship during its years of power consolidation.

The Nazi regime made use of both traditional and modern forms of self-celebration and self-justification, including the blending of technology with ideology.[25] The Autobahn and the architectural structures of Nazi Germany spoke for themselves, and the printed word and the poster enhanced their impact.[26] Visual media—including a technologically innovative use of film—and the printed word worked in tandem to support a predefined world-view. The German highway system, for example, required a textual and pictorial emphasis on the *Volksgemeinschaft* of its construction workers. In the Nazi scheme of things, this "community of individuals" was a racial construct wherein individual nature was subsumed into the Aryan mass working for the recovery of Germany.

The beauty of the airship as a masterful combination of aesthetics and aerodynamics also required textual direction, for the machine was no longer acceptable as a symbol of individual escapism. The Nazis intended to turn it into an instrument of fascination and submission. In May 1933, Propaganda Minister Goebbels used the airship to travel to Rome for his first meeting with the

Italian Fascists.[27] That fall, one diplomat suggested making it available at no cost to a League of Nations South American fact-finding mission, while a Württemberg official urged using it to strengthen links with Germans abroad "in the face of increasing Jewish propaganda."[28] Unlike other symbols of "decadent Weimar," the airship, with certain limitations, could be turned to the purpose of celebrating the rebirth of the nationalist state.

As part of the new *Volksgemeinschaft*, the Nazi party promoted a new mood of "airmindedness":[29] the nation was to prepare for air war on both defensive and offensive levels. This left little scope for the airship as a symbol of peace. Such limitations, however, were not yet obvious at the outset of the Third Reich, when both Propaganda Minister Goebbels and Air Minister Göring offered funding.[30] As the aircraft manufacturer Ernst Heinkel wrote, the airship, along with other German civil-aviation successes, offered a special "advertising moment" for the German technological spirit and will to survive.[31]

Such messages accompanied airship trips over Germany. On 1 May 1933 the *Graf Zeppelin* proceeded to Berlin in accordance with Goebbels's orders to attend May Day celebrations. The passengers invited on board symbolized the new regime's agenda for the day, a combination of German brotherhood and *Gleichschaltung* (coordination, or "bringing into the party line"):[32]

> Today we have again witnessed the wonderful result of German *Volksgemeinschaft* and of united popular will. The logbook gives us a clear sign of this. Next to the general stands the unskilled worker from Lake Constance, next to the son of the postal official, the deputy switchman, and next to the airship captain, the salesman. A single feeling fills us: gratitude to the unknown German worker whom we celebrate today, and who helped create this wonder of German technology. . . . [Down below] there is hardly a street in which someone is not parading. The main arteries are empty of streetcars and automobiles and filled with men, men, men, who are all moving toward one goal, who have but one wish, to be part of German work celebrations.

The emphasis on equality and national unity among the people inside and outside the airship is strongly reminiscent of a similar situation thirty years before: in effect the Nazi party was seeking to recreate through political manipulation what the crash of Count Zeppelin's LZ 4 at Echterdingen had given spontaneous rise to in 1908. Eckener had tried this, with limited success, during the Weimar era. In 1933, however, it was possible to choreograph a mass meeting of epic proportions to serve specific political needs; spontaneity had given way

The view from below: Hitler youth look up at the *Graf Zeppelin* in a prelude to Hitler's arrival at the rally on 3 September 1933. New Air Ministry rules required all German aircraft to have swastikas painted on the left side and the old imperial red-white-black colors on the right. In 1934 the rules changed to require that the Nazi emblem be displayed on both sides, reflecting the dictatorship's successful consolidation of power. *Völkischer Beobachter,* 5 September 1933, via ASTRA

to figuration or stand-in.[33] By removing individuality, the Nazi party sought to suggest that everyone imputed the same symbolic meaning to an icon, be it Führer, flag, or flying machine.

The airship's iconic capital did not just assist in this "anaesthetization of the masses" but also came to the fore in other ways. The nationalist upsurge became a pretense for selling new Zeppelin kitsch such as a cigarette card collection and photo albums that "every German" should own.[34] The stressed cultural value soon acquired a martial dimension, as became clear at the Nuremberg rally in September 1933.

The annual party congress, blessed with "Hitler weather," met on Zeppelin Field.[35] It included a choreographed propaganda show of countless youth parades and other displays among some fifteen thousand flags. The airship played an important role as the crowning element of the event leading up to the

Führer's appearance. Descriptions of the airship's arrival illustrate well the recognition and sympathy this generation-old symbol drew:

> For the deep droning noise of air squadrons cruising over the stadium a deeper, stronger one substitutes itself. Thunderously, *Graf Zeppelin* traces his majestic route over Germany's youth. The swastikas shine on the tail rudders. A frenetic jubilation shakes the whole stadium. Over one hundred thousand voices ring as a single immense scream of joy and pride at the giant of the air. He cruises slowly, shining in matte silver light over Nuremberg, returns, dips his bow to salute. A white floating sea of waving handkerchiefs responds from the Earth. Above, the giant lowers his nose again in salute, in thanks for the orchestra's play, for the exemplary enthusiasm that one hundred thousand hearts extend to him from below.[36]

The description of the unified male response to the airship and of its thunderous noise stands in marked contrast to those of the "good-natured giant" that had visited the Rhineland in 1930 following the departure of the French occupation troops. The airship's human occupants were no longer important. "He" was now a Baal-Moloch of sorts, an immense machine that, while suiting the modernist aesthetic of the Weimar era, also embodied the Nazi regime's preoccupation with gigantism.[37] The visual impact, one of a series of huge theatrical effects, formed a prelude to what was supposed to be the crowning event of the day, Hitler's essentially banal speech.[38] Ironically, he thought little of the airship.

Indeed, Hitler heartily disliked the dirigible, which witnesses said he feared as unsafe and too fragile. Comparing it to a floor varnish that had but one disadvantage—no one could step on the floor afterward—he once told several rocket scientists, "No, gentlemen, I will never sit in an airship!"[39] He even left the airship Autobahn survey, an important propaganda tool, to his work minister.[40] The airship certainly did not live up to Hitler's image of technology as a means to achieve personally controlled heroism. Hitler's personal architect, Albert Speer, recalled that the Führer had similar reactions to the jet-fighter. To Hitler's way of thinking, the airship was slow and awkward, the jet fighter powerful and dangerous, and a single man had only limited direct control over either.[41] Indeed, Hitler never flew on any airship, preferring the Lufthansa and Luftwaffe propeller airplanes specially converted for his use.

Despite speculation in numerous contemporary articles and in secondary literature, there is little evidence that the new Zeppelin dirigible paid for by the Nazi government was to be christened *Adolf Hitler.*[42] The Führer approved the christening of the new LZ 129 *Hindenburg,* as suggested by Eckener and former chancellor Heinrich Brüning.[43] He also preferred that name to the proposed *Germania,* which was the name of the Catholic Center party's newspaper. *Hindenburg* fit well with the Führer's choice of statesmen's and military commanders' names for other technological marvels, such as the battleships *Bismarck, Tirpitz,* and *Scharnhorst.*[44] Such names were also a politically convenient way to stress historic continuities between pre-Weimar Germany and the Third Reich.

The Nazis were determined to exercise complete control over Germany's cultural and national symbols, from rules governing the display of the swastika to the singing of religious and patriotic hymns.[45] The Air and Propaganda ministries were both involved in carrying out the instrumentalization of the airship as an icon. The Propaganda Ministry was required to submit requests to the Air Office for prior approval before confirming any flight plans. Everything from financial issues to the lists of guests flying on board the airship became a source of bickering between the two ministries.[46] In addition to overuse of the emblem,[47] this state of affairs may have actually limited the airship's potential as an instrument of Nazi propaganda.

The Nazis had no interest in the airship as an embodiment of individual creativity—a common association until then—but they still had to contend with this aspect of the mass cultural symbol when disseminating their ideology.[48] The Zeppelin would join a controlled collection of signs, all carefully selected and designed, and "fly happily under the swastika banner."[49] Such a collection would be a particularly effective way to proclaim the existence of a single Germanic culture, identifiable as such, the world over.

Once construction of the *Hindenburg* was completed, the Propaganda and Air ministries agreed on using it and the *Graf Zeppelin* for propaganda purposes during the 1936 elections. These were but a travesty of democratic referendum, but in the eyes of the Nazis such exercises were a form of further self-legitimization; style became substance. The airships' role as emblems of advanced technology would further stress *Volksgemeinschaft.* The preannounced flights were linked to national radio coverage and filmed for the German newsreels, to be shown in movie theaters throughout the country.[50] Radio logistics involved contacting all German districts for lists of personalities who would be interviewed from the airship. Operators would play background music during each broadcast of the "Conversations to and from the Airship."[51]

The content of these "conversations," however, required precise control. Coverage would include, for instance, the new Munich skyline, the latest Autobahn portions opened, industrial achievements, and National Socialist thought.[52] Since this was a technology-oriented show, speakers were instructed to avoid any countryside descriptions as banal. In effect, there was no place on the National Socialist airwaves for anything that resembled the idyllic commentary broadcast from the *Graf Zeppelin* in 1928 by Reichstag president Paul Löbe and others. Technology could serve only Nazi-approved ideology.

A takeoff incident caused a few delays (and was hidden from the public),[53] but the *Hindenburg* eventually joined the *Graf Zeppelin* in the air over Germany, cruising over Stuttgart and establishing a radio link with the Mercedes factory, which had manufactured the airship's engines.[54] Radio reporters on board received Godspeed telegrams and read them on the air. The *Hindenburg* tank regiment sent greetings; so did the Bremer Lloyd and Hapag shipping concerns and, of course, the Führer, then visiting the Tannenberg battlefield monument. Each overflown region offered an opportunity for historical or political commentary. Flying off Helgoland, radio speakers emphasized the island's long aviation history. Neither Danzig nor the Rhineland was forgotten; the airships dropped "memory leaflets" over these areas. The masses present in all major and smaller cities were clearly responding in an orchestrated manner. The disciplined reaction, which the Propaganda Ministry choreographed, called for flag waving at one moment, followed by attentive listening to the playing of the Nazi party's anthem, the *Horst Wessel Lied*. Throughout the three-day operation, loudspeakers on board communicated election slogans to the assembled audiences and on the day of the referendum called all "late sleepers to their duty." Zeppelin captain Albert Sammt remembered that propaganda officials continued mindlessly to issue their call even over the Essen cemetery.[55]

The airships also played a role during the referendum itself. Cameras on board the *Hindenburg* filmed an old lady from Friedrichshafen voting alongside airship captain Ernst Lehmann. The official voiceover noted, "One hundred four voters, one hundred four votes, one hundred four for the Führer. No one expected anything else."[56] The airships' ballet was portrayed as an overall resounding success: "Adolf Hitler's two helpers in the fight for Freedom and Peace"[57] became the archetypes of German machinery in the service of the new Germany.

Behind the scenes, Hugo Eckener's losing battle with the Nazis continued. Eckener had been a thorn in their side ever since they came to power. A speech he was asked to prepare in 1934 to support the Führer's succession of Hindenburg was rejected by the Propaganda Ministry and never broadcast.[58] His public joke at a society dinner that there had been two more yes votes than souls on board the *Hindenburg* hinted at his dislike of Nazi methods.[59] Not only had he opposed the election flights of the two airships on both political and commercial grounds—a previously scheduled transatlantic flight risked being delayed as a result—but he refused to board either machine.[60] As a result, he incurred the wrath of Propaganda Minister Goebbels, who issued a secret order forbidding the mention of Eckener's name in the German press on the grounds that he had placed himself "outside of the *Volksgemeinschaft*."[61] The order, however, became an open secret in Berlin, and it backfired abroad, where newspaper editorials criticized the decision.[62] The incident clearly demonstrated the Nazis' growing unease regarding both the airship and Eckener.

It was difficult to separate the two. Eckener was a cultural icon solidly in place by the time the Nazis came to power, and his popularity made it hard for them to manipulate the Zeppelin. To sympathetic Americans, he had become "the embodiment of the best German national characteristics."[63] He commanded respect abroad, even in France. The first German to win the annual medal of the Fédération aéronautique internationale, he also received the gold medal of the Royal Aeronautical Society. The reputation he had built during the Weimar era as a tireless speaker enhanced his charisma and popularity now. He alone could maintain the airship's image, although others could probably replace him in other technical and business functions.

The Zeppelin company, of course, faced *Gleichschaltung* as other businesses had done. Machinists automatically became members of the German Work Front, and engineers participated in aeronautical and technical colloquia through other Nazified professional organizations.[64] Some even joined the party, although their motives are unclear. In the words of one Zeppelin contemporary, several "young lions," especially among the navigators, became members, while "older men, who thought little of Nazism in 1934, apparently changed their thinking by 1936. Witterman [sic] is an example."[65] Anton Wittemann was but one of two out of seven active airship captains to have joined the party by the time of the *Hindenburg* disaster (some may have expressed open sympathy for the regime without joining).[66] To Nazi officials, however, controlling the company meant additional opportunities to manipulate the airship's image through the Nazi interpretation of "airmindedness."

Although presented as a German trait, airmindedness existed elsewhere.

Western democracies practiced it in their own ways, celebrating the heroic raids of civilian fliers as an affirmation of the nation's place in the modern world and its superiority to colonized peoples. Other manifestations of airmindedness reflected contemporary concerns: in Switzerland, blueprints for civilian air defense were implemented in 1935, while in France, the Popular Front of 1936–38 subsidized flying lessons and "people's air clubs."[67] In most cases, such initiatives depended on individual willingness to participate, yet the multitude of expressions proved that the fascination with aviation was general among industrialized nations.

In dictatorial systems (the Soviet Union and Italy), aviation had been used to help orchestrate social control since the 1920s.[68] The Nazi approach to mass mobilization, then, while relying on Versailles revanchism, did not offer anything new. As a dictatorship, the Nazi regime could easily channel popular enthusiasm away from individual endeavors reserved for regime heroes, and prioritize strictly air-defense issues and training. Thus, gliding, a favorite escapist hobby during the Weimar era, now became a "discipline sport."[69] The airship still fit quite well into this rigorous program. What better way to promote gliders than by launching one from the *Graf Zeppelin?* The German Airport League was able to secure a slot in the airship's strictly controlled propaganda schedule to proceed with an experimental drop carried out by Wiegmeyer, one of Germany's elite glider fliers. Although this kind of operation was never repeated,[70] it served to demonstrate the apparent union between two pioneering flight techniques, Count Zeppelin's and Otto Lilienthal's. Press coverage of this form of air education was extensive.[71] Such publicity stunts served the Nazi propaganda machine well, and other examples followed.

For the first anniversary of the Nazi coming to power, the postal service issued aeronautical stamps that included portraits of Count Zeppelin and Otto Lilienthal and an image of the Nazi eagle diving toward a globe.[72] The undertaking reflected Germany's desire to free itself of the Versailles treaty restrictions and to honor its aeronautical antecedents. The ultimate aspect of airmindedness, however, revolved around education.

As of December 1934, the Nazi school curriculum included a new section devoted to aviation and its technical, scientific, and geopolitical implications. Lesson plan recommendations included using the "Zip Zap Zeppelin" poems of imperial times and a blackboard drawing of an airship (complete with swastikas) to familiarize children with the rudiments of writing and drawing and with the story of the proud old count and his achievement.[73] Older students would learn about meteorology and model building. Most youth publications, although meant to reflect *völkisch* values of blood and soil, were to incorporate modern German achievements in industry and technology.[74] In the case of aviation, although airplanes and gliders got the lion's share, texts on "Germany's renewal" always contained a section on the count and the airships that may have enchanted these children's parents. Whatever its implications, the airship (and its maker) belonged to the cultural heritage and required acknowledgment.

Technology would also justify and strengthen the Nazi view of history.[75] To this end it was necessary that Germans adopt the National Socialist position on the legacy and value of technical feats.[76] Just as the concept of airmindedness was to introduce an understanding of air warfare through education, so carefully tailored popular accounts of great German men and technological successes would produce pride in the official version of the past as well as enthusiasm for the supposedly unique modernist stance of the state.

A German inventor's technological success was an ideal means of supporting the new world-view. The Nazi ideologue Alfred Rosenberg attempted to link the role of technology to the Germanic myths. "Nordic Greece's" dream of Icarus led directly to the mythical Wieland's aching for flight, and to the very real invention of the combustion engine, which in turn allowed Count Zeppelin to make the dream a reality. Such triumph was proof, in Rosenberg's view, that technical thought was a new expression of will that would unite the German people and render it independent of any capitalist intrigue.[77]

Other comments on the count echo this theme. Among the historical works written for public consumption during the Nazi era, several included Count Zeppelin in a series of "Great German Men." What came out, though, was a decidedly Nazified personality. This was a different image from either the proud, principled soldier or the gentle old man the previous generation had come to picture. In these new accounts, he was presented not as the persevering patriarch but as a fighter of the likes of Admiral Tirpitz. Although this emphasis was not inaccurate, the suggestion that Count Zeppelin saw the ultimate triumph of flight strictly in terms of combat—many quoted his 1916 claim that Germany stood in "a fight for existence"—and himself as a sort of Führer of technology owed much to the post-1933 *Zeitgeist*.[78] As the lone hero, the count had succeeded in launching a new cultural era that the Great War ace Ernst Udet summed up thus: "One German soldier conquered the sky, and German soldiers will master it!"[79]

Yet this was only half of the Nazi equation. A Führer had to lead a people. In these accounts, the people are acknowledged as a force to be reckoned with, yet they appear passive, standing behind the great man in a manner reminiscent of the Hitler cult and strikingly unlike the varied reactions the count inspired in his own time. Perhaps for this reason, some accounts of the life of Count Zeppelin found it convenient to end in 1909, since after that time the count could no longer be considered the lone knight fighting for German honor.[80] By then, the *Volk* had intervened through the spontaneous fundraising drive.

The memory of the count was exploited in other ways as well, to instill in German youth a culture of endeavor and innovation. These and other popular accounts of the lives of great men did not specifically distinguish the Nazis or Germany from other regimes and nations, but they did emphasize the general ethnic superiority of both German men and machines. Such accounts echoed Georg Hacker, one of the first Zeppelin captains, who emphasized in his memoirs that airships built abroad "were missing the German experience, the tradition of Manzell and Friedrichshafen. The Zeppelins appear to have been reserved for the German people."[81] The 1935 crash of the American-built *Macon* airship was cited as confirmation of such views.[82] Similarly, a children's book described the airship as a machine proper to this nation alone: though built of materials coming from all around the world, it became a technological entity only through German efforts.[83] The technological culture described in such works was distinctly male, despite the cross-gender appeal of Count Zeppelin and his machines, but since the airship community itself was almost exclusively male, it fit well into the Nazis' concept of the proper division of labor in society.

Viewed in this light, Count Zeppelin seemed to have a clear place in the Nazi pantheon—a cult of dead heroes that combined myth and reality under the emblem of the integral national state.[84] Not only was Ferdinand Otto Heinrich Graf von Zeppelin a "full-blooded" German whose genealogical roots went back at least nine generations into sixteenth-century northern Germany,[85] but his aeronautical endeavors fulfilled the Nazi requirement that Germany aspire to a future in the air. The count's childhood was described as "studious and soldierly," in notable contrast to the effeminate Christian world-view he was said to have been exposed to.[86] His military spirit also came to the fore in his 1863 trip to the United States and his gallantry during the Franco-Prussian War.[87] His pursuit of his goal had demonstrated strength of character, that combination of relentlessness (*Unerbittlichkeit*) and flexibility that characterized the warrior ethos. These soldierly qualities, in Deputy Air Minister Erhard Milch's view, were what had made Count Zeppelin the most popular truly German (*kerndeutsche*) figure in Germany.[88] As if to drive the point home, in 1938, the centennial of the count's birth, Germany's first aircraft carrier was named after him in a spirit of tradition-consciousness meant to commemorate the "holy power" that had conquered airspace for the glory of Germany.[89]

Each commemoration of the count's memory, from the August 1933 jubilee of the Echterdingen crash to the twentieth anniversary of his death and the centennial of his birth, emphasized his mastery of destiny through technology.[90] Only Hugo Eckener's biography of Count Zeppelin strove to provide a more balanced account, although it too suffered from distortions apparently imposed by the publisher.[91] Even during World War II, long after the end of the airship era, the count continued to be revered as a true "Aryan," honored upon the anniversary of his death alongside the Nazi martyr Horst Wessel. In addition, literary celebrations of both the count and his machines were all given substantial space in new exhibits set up around Germany, including a Zeppelin museum in Friedrichshafen.[92]

Zeppelin propaganda was thus a central, yet problematic, element of airship activities until 1939. It also relied on the notion of commercial enterprise—the element Eckener had stressed throughout the decline of the Weimar Republic as he desperately sought new funds.

Airship Travel Comes of Age

Like most other companies in Germany, Zeppelin suffered tremendously from the world economic crisis of the 1930s. The company had hoped that the airship's recent successes as a means of transport would attract foreign investment as an alternative to the limited funding available in Germany, but little came of these prospects. An American-German joint venture calling for regular transatlantic service fell through following the Wall Street crash, and Spain's plan for building a new airship field was held back by both political and financial unrest.

In fact, the Zeppelin company was coming to depend more and more on government handouts. Airship flights overseas had failed to provide enough income to cover expenses. Zeppelin visits to South America had begun in 1931, but limited funding prevented a wider development of infrastructures. The airship had to use its own base in Friedrichshafen, a region far from major industrial and financial centers. Eckener now began to argue for the opening of an airship port in Frankfurt am Main, claiming that it would serve both culture and business (few transport operations, he said, could claim having covered distances of 1.5 million kilometers in four years).[93]

Although the Nazis cared little for Eckener's convictions and regarded the airship as primarily a propaganda instrument, they were willing to set aside their peculiar economic principles in the case of the Zeppelin company. Unlike other industries that the Nazi party relied on to consolidate its economic power, the Zeppelin company had little to offer in the way of material advantage. A massive injection of capital would be required before any real financial returns might register. It was therefore logical to try to turn the airship into a profitable venture, even if the plan did contradict the Nazi view that technology was more properly the handmaiden of culture than of business.

In addition to a 3-million-mark loan and funding from the state of Württemberg to build a new hangar, the Zeppelin company received some 900,000 marks that were earmarked for the development of its South American route.[94] Further evidence of governmental support appears in a 1934 document that deemed the value of domestic airship transport to be too low and recommended expansion overseas through a kind of German monopoly. The document expressed the hope that such an endeavor would broaden the outlook of National Socialism in technocultural affairs, commerce, and transport.[95]

In 1935, while the new LZ 129 airship was still under construction, a new company separate from the construction firm and called Deutsche Zeppelin Reederei was founded. DZR provided the Nazi government with a way to control the Zeppelin company without constantly having to deal with Hugo Eckener. The new business firm may also have been the outcome of competition between the Air Ministry and the Propaganda Ministry. Propaganda Minister Goebbels had offered the Zeppelin company some 2 million marks to complete a new airship in the summer of 1934.[96] The Air Ministry responded with an altogether grander plan: 9 million marks to fund an entirely new "airship line" and provide a stable capital base for operations.[97] Lufthansa, itself under government subsidy, held part of the shares of the new company. On the occasion of the company's inauguration in March 1935, Air Minister Hermann Göring, who despised airships, nonetheless chose to emphasize the Zeppelin's value even in the face of the airplane's "great position." Göring also noted that the absence of airship crashes in the Fatherland since World War I proved the quality of German engineering, workmanship, and enterprise.[98]

As for the new firm, its composition and close ties with the German government were seen as a guarantee of early victory in the air for the "new" Germany. Eckener became chairman of the board, a position he apparently accepted because it still left him with consultative powers. The executive officers represented Lufthansa and the Air Ministry.[99] Press releases stressed the war experience of the two new company managers. Both had fought valiantly in the Great War—good preparation for the struggles of business. To Eckener, this meant that he would have to say yes to them more often than no.[100] None of Count Zeppelin's descendants were invited to be on the board, however, the official reason being that none of them were men.[101] It was essential to keep technology a masculine endeavor, even though the airship belonged to the German *Volk* as a whole.

In line with DZR's businesslike operating routine, the emphasis in its press releases was less on cultural iconography and more on German commerce. The *Graf Zeppelin*'s transatlantic crossing, with its on-time arrival, was a way "to bring to the world the value of German work."[102] Evolving into a "symbol of the German *Volk*'s peaceful determination to work toward reconstruction,"[103] the airship maintained an edge in international competition, fighting for German business and jobs. Nazi ideology demanded living space for the *Volk,* and aviation would help it expand into new territories; as for the Zeppelin path, it was to be exploited further as a kind of ersatz colonization, supplementing autarkic programs.[104] Technology did not just have a cultural role to play but was also part of the world of business and profit, with all those aspects coming together under the single emblem of the airship. The theme was far from new, for both Count Zeppelin and Eckener had sought to establish a multifaceted image for the airship, but now that image was entirely under governmental control.

Meanwhile, at the Friedrichshafen factory, the *Hindenburg* was nearing completion. Only superlative terms could describe its mass, some twenty times greater than LZ 1's. It also represented the zenith of airship design.

The *Hindenburg*'s hull number was LZ 129. The gap in the sequence—the *Graf Zeppelin* was LZ 127—reveals a shift in engineering decision-making. By 1928, Zeppelin engineers had begun drawing up plans for a new airship, the same length as the *Graf* but greater in diameter. The propulsion system would be similar to that of its immediate predecessor, but it would carry passengers within its hull rather than in an area behind the control cabin. Plans proceeded apace, including the erection of a shed capable of housing construction of a wider airship. But then a British airship crash led to a reconfiguration of Zeppelin designs.

In Britain, the government had sponsored the development of a rigid-airship program for use in linking the colonies to England. A combination of political manipulation, competition between two separate airship programs, and insufficient testing led to the rushed decision to fly one airship, R 101, to India in October 1930 as a show of British technology and prestige. The airship encountered a heavy storm soon after leaving England, and it crashed and exploded in Beauvais, France, killing forty-eight of its fifty-six passengers, including the British air minister.[105] The disaster signaled the end of British involvement in airships, but it also prompted a complete redesign of the LZ 128 project.

Soon after receiving news of the R 101 tragedy, Zeppelin's chief engineer Ludwig Dürr, in consultation with his team, scrapped the outlines of LZ 128 and had the airship redrawn to contain nonflammable helium instead of hydrogen. Since helium was slightly heavier than hydrogen, a greater volume of gas would be needed to carry the same number of passengers. The new design abandoned the lift- and propulsion-gas combination that characterized the *Graf Zeppelin* and called for the use of far safer diesel engines.

Fortunately for the Zeppelin company, the rearmament program that followed quickly upon the Nazi assumption of power offered the ideal engine. The Daimler-Benz corporation had completed a powerful diesel engine for use on torpedo boats that, once modified, perfectly suited the needs of an airship. This meant losing the advantages of Blau gas, the propellant arrangement that allowed the *Graf* to maintain weight as fuel was spent, but engineers eventually devised a method for recovering water, which would replace the diesel fuel as it was consumed. The water would then serve as ballast and be flushed when necessary.[106] Helium, however, could not be obtained. Available only in the United States, it required a special export permit. Despite pleas from Eckener, the U.S. government declined to authorize the supply of helium. In the end, then, the *Hindenburg* flew on hydrogen.

The new machine had all the features of elegant modernity. The control car was much smaller than that of the *Graf Zeppelin* because the passenger cabin was tucked into the hull's belly. The idea followed the patented 1919 designs of Paul Jaray, who had stressed the streamlining advantages of such a solution.[107] Other novelties included a specially built Blüthner aluminum piano[108] and design features inspired by the Bauhaus, an institution that, paradoxically, the Nazis had already outlawed. In tune with the times—and as a response to passenger complaints from earlier Zeppelin flights—the design team arranged

LZ 129 under construction. Taken in February 1934, this picture shows the laying of the ring sections. First assembled flat, each ring was then hoisted and mated to the fuselage. Wiring then followed. ASTRA

to incorporate a smoking lounge, which became, in the words of one experienced Zeppelin passenger, "unquestionably the most popular place on the airship."[109]

However remarkable the *Hindenburg* was for the originality of its engineering and the elegance of its design, its size was what got the attention of the Nazi-controlled press. The media reported on the construction of the monster as if it were some kind of superhuman feat, another illustration of the *Volksgemeinschaft* in action. Airship construction was indeed a stupendous undertaking, though it did not quite live up to the impression of excitement and energy that columnists sought to communicate. The work of an airship technician lay midway between artisanship and industrial engineering. Meticulous endeavor, at times extremely slow, characterized the factory. Media accounts and artistic drawings ignored such realities, depicting the Zeppelin factory as a frenzied plant, thereby matching the dynamic stereotype of the National Socialist workplace. There, each technician was described as rushing about his task, checking on nuts and bolts, "for each small part, no matter how insignificant, has its purpose and goal," in the words of one press correspondent.[110] The analogy to the wider conception of the Nazi state is unmistakable. It points to the basis for industrial war preparation, but also to the atomization of the individual in the face of the greater good,[111] and represents the next stage of the process of *Gleichschaltung* and *Volksgemeinschaft*, coordination and community, imposed from above.

The machine's completion was delayed several months by erroneous assembly schedules, but this did not dampen spirits.[112] Even the traditional debate about the technical and economic merits of the airship relative to the airplane was glossed over: "To us, it is not airplane or airship, but rather airplane and airship."[113] In fact, however, by 1936 the airship was losing ground to the airplane, as Eckener himself acknowledged.[114] Several airplanes were capable, with refueling stops, of beating the airship to a long-distance destination. Nevertheless, the Zeppelin remained a serious contender for long-distance travel thanks to its combination of range, sophistication, and standards of comfort (however exaggerated the claims for those standards may have been).[115] The airship was the epitome of elite travel, "more comfortable than a first-class sleeping coach."[116] The lucky few who could afford to board the mythical machine could attempt through their own "heroic" expedition to beat Jules Verne's character Phileas Fogg in circumnavigating the globe.[117]

With two airships in service and a third in the planning stages, the Zeppelin

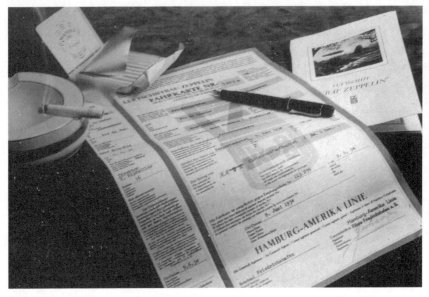

Zeppelin luxury. The details of this publicity still—cigarette case, travel guide, and fountain pen framing a Zeppelin ticket—suggest the epitome of luxurious travel. The high cost of a ticket limited airship travel to a select few. The Hamburg-Amerika shipping line acted as an agent for the Zeppelin company in Germany and abroad. LZA, 127-0454

company sought to make airship travel a routine matter. However, each trip had its own unique details, as passengers' memoirs show. These ranged from the expected personalities of the day coming on board to a series of "firsts," such as Father Paul Schulte celebrating the first airborne mass (to the dismay of the Nazi party).[118] Some of these events provided rich fodder for the media, such as the encounter between the *Graf Zeppelin* and the German ship *Monta Rosa* over the South Atlantic, where the two crews traded a bottle of champagne.[119]

Generally, airship passengers first flew in from Berlin on a Lufthansa plane, then stayed in Friedrichshafen's Kurgarten, enjoying that "small spot of German soil" before boarding the "luxury flying hotel."[120] Standing next to a bust of Field Marshall Hindenburg, a steward greeted visitors and collected all matchbooks and lighters. Passengers were then led to their cabins, shown the bathroom facilities (shared, though each cabin was fitted with a wash basin), and lectured on the three rules: no smoking outside the specially designed smoking room, no unaccompanied strolls outside passenger quarters, and no throwing of objects out the windows. This last rule stemmed less from ecological concerns than from the fact that the wake of the airship behaved much like that of a seafaring ship: discarded objects flew back toward the hull or into the propellers, damaging them.[121] When all was set for departure, the liftoff rituals began, as hands released the moorings and the machine rose. Takeoff ceremonies often included a local band playing the German national anthem. Whatever the rituals, the point was to offer a balance between routine and the special status of airship travel.

Airship travel literature was heavily directed at foreigners. Brochures printed in English explained how simple and fascinating travel really was. While each passenger's luggage allowance was limited to 30 kilos, potential clients were assured that all amenities would be available on board, from telegram dispatch to games, cigarettes, and of course alcohol. Marketers made a special point of addressing the currency issue. Nazi currency regulations were so strict that the use of foreign money could become problematic even on board an airship. Consequently, fares were all-inclusive, and passengers were encouraged to set up a small credit account to cover incidentals and avoid the nightmare of currency conversion.[122]

The experience of travel itself was, of course, central to the selling of the airship experience. Siegfried Kracauer's observation about the new class of consumer-travelers—that they were "playfully excited about the new velocity, the relaxed roaming about, the overviews of geographic regions that previously could not be seen with such scope"—is remarkably pertinent to the airship's passengers.[123] Whether observing the passing of an ocean liner or gazing at the distant landscape, its occupants could enjoy the distinct advantage of airship technology. With nothing to fear from either sea- or air-sickness (the company proudly reported that none of its clients had ever experienced motion discomfort), passengers benefited from the machine's remarkable stability. By all accounts, the noise of the engines, placed well behind the travel quarters, was but a distant droning. Compared with the storms that the *Graf Zeppelin* had suffered through, the squalls the *Hindenburg* encountered did not make much of an impression. Nonetheless, ocean-liner passengers still enjoyed better conditions. Even with first-class treatment, airship sleeping births were cramped, and water-supply limitations required passengers to content themselves with shower baths.

The airship may have been lacking in certain comforts for the traveler, but its appeal to business interests appeared to be unqualified, or so the advertising suggested. The *Hindenburg*'s first South American flight carried the five hundred thousandth Opel car for delivery to Rio de Janeiro: the German reputation sold German goods. A poster for the 1937 season represented the *Hindenburg* hovering over the Empire State Building and promised, "Two days to North America": German engineering was seemingly taking on the supreme symbol of the American technological sublime, and therefore of American business.[124] Even Eckener played along with the theme, permitting pictures of himself in a Lord Kitchener–like pose to appear in American and German ads urging readers to buy German products.[125] The campaign had some success. By the end of 1936, joint ventures with the United States and Great Britain had reentered the picture, and three more *Hindenburg* sister ships were planned for completion by 1939.[126]

Profit was indeed a necessary and welcome implication of the technocultural link, and a reminder of the compromises that had to be made if one wanted to succeed. Eckener's background as an academic who had become versed in technology and business, as well as his current position between an economic rock and a political hard place, epitomized the dilemma that many German professionals were experiencing. His expertise and image remained essential to the running of the Zeppelin company, despite the tensions this created with Nazi doctrine.[127] His appearance on posters was nothing more than a deal to which he could agree as a businessman despite his opposition to autarkic policies. The message from the picture, however, was that a promoter of German technology had, like other professionals, become a "soldier of the Reich."[128]

Naturally enough, German advertising abroad that had to do with technological achievements also presented the airship in terms of the theme of keeping "Germans united abroad."[129] But while the Nazis' version of the airship as an emblem of nationalist *Gleichschaltung* may have influenced some expatriate Germans, it proved counterproductive in Switzerland.

Since 1908, the appearance of Zeppelins over Switzerland had hardly ever caused friction. The Zeppelin company ran regular charter flights over Swiss territory, taking passengers over mountain chains and the idyllic Rhinefalls at Schaffhausen. When regular transatlantic service began, the Swiss government granted monthly and even annual overflight permits.[130] There were occasional incidents involving passage over restricted zones, but the disputes were always settled quickly.[131]

With the Nazi takeover, however, things changed, and by 1937 tensions had increased markedly. In April of that year, despite official warnings, the *Hindenburg* had overflown Swiss military maneuver areas. Official reports noted that the militia soldiers were quite angry about it and that civilians in both

German- and French-speaking areas were fed up with the overflights. Furthermore, the Swiss government seriously considered bringing a charge of espionage, a radical departure from its hitherto appeasing stance.[132] The novelty appeal of the airship was wearing thin, and calls for sanctions came in from all sides of the Swiss press. The German press in turn attacked Swiss newspapers for denouncing the Zeppelin's activity, an obvious warning that the Reich authorities would not take criticism lightly.[133] A Swiss ban on charter flights to the Rhinefalls would likely mean retaliation from the German Air Ministry, perhaps in the form of a barring of Swissair flights to Germany.

Eventually the Swiss decided to issue a formal protest and require advance notice prior to any Zeppelin flight. This would give some satisfaction to a public that saw little benefit in the regular disturbance of Swiss airspace by a flying German symbol.[134] A machine that for almost thirty years had delighted Switzerland now became the focus of its frustrations in an era of economic troubles and political doubt. True, Nazi Germany had sympathizers among its southern neighbor's people, but such understanding was swamped by the ongoing tension and by the incidents local Nazi groups were causing.[135] The airship had now become associated with such troubles.

Helium, Hydrogen, and Humble Endings

German-Swiss tensions over the airship disappeared, however, when catastrophe occurred. On 6 May 1937, following its first transatlantic crossing of the year, the *Hindenburg* came in to land at Lakehurst, New Jersey. It suddenly caught fire and was destroyed, killing thirty-seven. Unlike other airship disasters, some of which had been captured in still photographs, this event was recorded on newsreels. The news captivated the world. The shock in Germany was of course great, and both the authorities and Eckener moved quickly to control it.

A shaken Eckener flew back from Austria. Upon arriving in Berlin, he learned that the *Hindenburg*'s captains, Max Pruss and Ernst Lehmann, both wounded in the disaster (Lehmann died soon after), had claimed in a state of shock that only sabotage could have caused the crash. Eckener too, when he learned of the crash in a late-night phone call from the press, had thought of sabotage. However, when he went on radio for a national address, he dismissed the notion as fantastic.[136] Eckener later explained that he could not conceive of sabotage occurring on U.S. soil, despite having received threats during the

1933 flight of the *Graf Zeppelin* to Chicago. The possibility of a weather-related problem was easier for him to accept, although Zeppelins had in the past been struck by lightning and flown on without trouble. While Eckener brushed aside all notions of sabotage, the rumors lingered on in Friedrichshafen, dividing Nazi sympathizers, convinced of conspiracy, from top management and apolitical employees, favoring the thesis of an accident.[137]

Outside the Zeppelin factory, however, people were preoccupied with mourning, not questions. Germans felt genuine grief at the tragedy, and editorials commented on it in terms that did not echo the standard propaganda. "The airship was more than technology," stated a Frankfurt newspaper; "it was a personification. . . . Children went to school sad and gloomy-faced . . . and have aged from the experience."[138] However, such remarks were soon drowned out by authoritarian rhetoric. Rabid antisemites claimed that "a Jew stands behind the explosion of our airship" (to some eyes, the cloud of smoke had apparently been shaped like a "Jewish" face).[139] Other Nazis held up the victims of the disaster as martyrs—the dead crew members were "soldiers"—in whose name the fight for Germany would carry on.[140] Following repatriation of the bodies, all the victims of the *Hindenburg* accident were given state funerals and eulogized as "joined with the spirit of Count Zeppelin."[141]

In many communities throughout Germany, spontaneous fund-raising drives began, as if memories of the 1908 crash at Echterdingen had kicked in to bring back the unified support Germans had then shown Count Zeppelin. Embassies abroad also reported receiving funds for the replacement of the airship and requested instructions on how best to handle the money.[142] The authorities quickly put a stop to these demonstrations of popular support. The Interior Ministry ordered the termination of all unauthorized fund drives. If a particular fund drive had not been authorized by the government (along the lines of the *Eintopf* food campaigns), it would not be tolerated, even as an expression of grief. The official explanation for the banning of such efforts was that the airship had been insured (which was true).[143] Yet the subject would not be closed so quickly. In fact, contributions continued to come in, many of them sent directly to the German Air Ministry, forcing a public announcement that a special account had been opened with the Reich Central Bank.[144]

Although the foreign contributions sent to help Germany cope with the loss of the *Hindenburg* were somewhat limited, they were noticeable enough to be mentioned in the German press. Switzerland, described as the land after Germany where the Zeppelin was most popular, displayed shock and dismay, while

Austria shared in this "stroke of fate" that had hit not just the German Reich but all German people.[145] The funds collected abroad were officially routed to support the "German House of Fliers," a senior airmen's home, and to build monuments to the *Hindenburg* in both Friedrichshafen and Frankfurt.[146]

The fund-raising episode is significant in several respects. For one, it is telling that the government did not capitalize on this spontaneous display of national solidarity but instead felt the need to stifle it. For another, it demonstrates that Zeppelinism as a social phenomenon remained strong, even without the full propaganda apparatus that had made the airship an official sign of Germany's greatness. At the official level, the decision on the fund drive also reflected uneasily on the future of the airship. Many of those concerned continued to hope that the promises Göring had made would become reality. Technological histories stating that the airplane had by now completely overtaken the dirigible overstate the case: contemporary projections suggested that it would be at least a decade before this happened, even weighing the high costs on the side of the dirigible.[147] The airship still had great potential, and the explosion at Lakehurst only confirmed what had been known all along: finding helium was a priority.

Originally planned for helium inflation, the *Hindenburg* ultimately had to be filled with hydrogen because the Germans could not obtain the other gas; the United States was the world's only producer of helium, and its export was banned under a 1927 act of Congress. The notion of helium inflation was therefore abandoned until disaster struck. There now began a frenzied search for helium to fill the new LZ 130, the *Hindenburg*'s sister ship still under construction. As negotiations proceeded through diplomatic and business channels in the hope of breaking the export ban,[148] a multitude of reports surfaced indicating the availability of helium elsewhere. From Rio de Janeiro to Turin and even within Germany, claims emerged pointing to the existence of helium.[149] Rumors began to circulate that foreign airlines were trying to stall the helium negotiations. Until Lakehurst, the dirigible had maintained a surprising record of efficiency and safety compared with airplane and hydroplane services.[150] The German Air Ministry was well aware of this, and of the "good reputation German airship transport has earned for years the world over." Should helium prove to be unobtainable, only then would Germany's reputation be at stake.[151] The Air Ministry, weary of the Americans' repeated refusal to grant an export license, even asked the Foreign Ministry to send out feelers on possible European sources of helium.[152]

LZ 130 under construction in June 1937. Begun in 1936, the sister ship of the *Hindenburg* first flew in September 1938. The installation of the fabric involved an intricate process of cutting, sewing, pasting, and doping the material to make it waterproof. ASTRA

Following the loss of the *Hindenburg*, the U.S. government had in fact decided to lift the ban on helium exports to Germany, but the decision was suddenly reversed in the spring of 1938, apparently in response to the *Anschluss* with Austria.[153] In April, Hugo Eckener, whose reputation abroad remained intact, set off by ship to the United States to try to persuade American officials to reverse their decision, but to no avail. Despite support from certain U.S. Navy officials and the press (which had criticized Commerce Secretary Harold Ickes, who alone could approve an export clearance), the decision stood.[154] Nevertheless, German ambassador Dieckhoff, reporting on the meeting between Eckener and President Franklin D. Roosevelt, concluded that Eckener's presence had saved face for Germany, and that the airshipman's personal friendship with the American president implied that all was not lost. For this reason too the German Propaganda Ministry muzzled the press and forbade any anti-American comments, in case American supporters of the German cause eventually got their way.[155]

In the end, the government's hopes were dashed: the United States chose not to provide the helium Germany required for its airships. That refusal was a major factor in the eventual decision to abandon the airship as one of the symbols of the "new" Germany. Yet the Zeppelin was not erased from popular memory. In fact, a poll conducted in September 1938 to test the historical and political indoctrination of some 259,000 workers showed that the inventor of the German rigid airship had one of the highest rates of recognition among various prominent German figures and events. Count Zeppelin was better remembered than the Teutonic hero Siegfried, the composer Beethoven, or even the years of the Berlin Olympics and the annexation of Austria. In some cases even the symbols that Nazi propaganda had incessantly evoked, such as the swastika, reached levels of recognition only roughly equal to those attained by the count.[156] Eckener did not appear in the questionnaire, although some "incorrect" responses mentioned him, alongside Göring and Hindenburg, as the father of the airship.

The polling results also confirmed that while active Nazi involvement in the airship business faded following the loss of the *Hindenburg*,[157] the airship could not be written out of the symbolism of the Third Reich. After all, Hermann Göring himself had promised after the disaster that the program would carry on, even though behind the scenes both Hitler and Goebbels wanted to give the airplane priority. The propaganda minister had eventually sided with his master: "The Führer is right," he wrote following the crash, "the future belongs

A publicity still of the lounge of the LZ 130 *Graf Zeppelin II*. The ship received the same name as LZ 127 because the latter was due to be retired. The lounge painting represents LZ 127 cruising over Rio de Janeiro. The Zeppelin company commissioned such photographs during the helium negotiations period, expecting that passenger service would soon resume. In fact, LZ 130 never carried any paying passengers. ASTRA

to the airplane."[158] As the board of DZR concluded in its 1939 report, popular interest remained strong enough that the Postal and Air ministries went on to authorize and sponsor demonstration and mail flights over several cities.[159] Even when the controlled media shifted the aeronautical focus toward the airplane (some historical summaries mentioned Lilienthal but forgot Count Zeppelin),[160] it still covered airship events as symbols of optimism.

In September 1938 the new LZ 130 *Graf Zeppelin II* flew as part of a medium-distance trial to the newly annexed Austria. The trip coincided with the Sudeten crisis, and the local press described the new masterpiece of "German engineering art" as a sign of hope for the fleeing German refugees. Pictures of the giant showed it hovering "majestic, shimmering, victorious" over Viennese passersby studying air-alarm posters.[161] Here, at the end of the airship era, the technological marvel was once more instrumentalized as a Nazi emblem of political success and military power: leaflets dropped during an airship flight over the recently annexed Sudetenland proclaimed to the local population, "For you the nation would have drawn its sword!"[162]

The airship continued to enjoy strong recognition, and numerous local Nazi leaders expected it to visit areas throughout Germany for purposes of entertainment and national celebration. The few flights undertaken in 1938 had translated into a booked calendar for 1939 and even 1940, a welcome state of affairs for DZR.[163] In addition, another steady source of income had materialized in the form of the old *Graf Zeppelin*. Emptied of hydrogen and parked in a hangar at Frankfurt Airport, the machine had become a de facto museum drawing hundreds of visitors eager to catch a glimpse of the Zeppelin legend.

However, the apparent status quo was under attack at the ministerial level. In late 1938, following the failure to obtain helium, the decision was made to discontinue the airship program (although it was agreed that it would be a blow to national morale to retire the *Graf Zeppelin II*). Eckener was chairman of the board of DZR but had little power compared with the representatives of Lufthansa and the Air Ministry, both of which supported the airplane. In any case, without helium there was little to do but sit and wait for better days. A possible sale of the newest airship to the Soviet Union was discussed, and even the establishment of a new Mideast link, but none of these projects came to fruition.[164] The one use eventually found for the *Graf Zeppelin II* was a military one: a secret flight off the British coast in August 1939 undertaken in an attempt to detect radar installations.[165] After that, however, the machine joined its sister ship at Frankfurt, where newspapers encouraged public visits on the grounds that such a show of interest would prove that "aeronautics is today the flesh and blood of every German."[166] In fact, things continued to seem normal in August 1939, when LZ 130's certificate of airworthiness was renewed for one year.[167]

It was not until the start of the war that the Nazi bureaucracy moved in to suppress the remnants of the airship age. Under the guise of "war needs," and despite ongoing negotiations on the part of the Zeppelin company, both dirigibles were declared obsolete. Efforts to save the machines and their sheds on the part of several personalities, including a few Nazi regional leaders, came to nothing. On 20 February 1940, the Air Ministry ordered the destruction of the airships and the transfer of all plans to one of its departments, along with certain pieces to be preserved for "teaching purposes."[168]

Eckener, now seventy-two, judged the debate to have been lost for the moment, but he hoped that retaining the Friedrichshafen installations might mean the possibility of starting over after the war.[169] A call for another *Spende* might work, after all. By now, however, the airplane was clearly forging ahead, beating the dirigible even on its own South American mail route. The final nail in the coffin came from the United States. The failure of DZR to prosecute a 1938 application to operate a new airship service offered new grounds to the Civil Aeronautics Board to rid itself of the issue three years later at a time when U.S.-German relations were deteriorating.[170]

Why did Nazi leaders want to see the machines destroyed rather than placed in a museum, which had been the fate of other white elephants such as the giant Dornier DO-X hydroplane? Recommendations from several high-ranking officers failed to alter the course of events. Captain Max Pruss, a survivor of the *Hindenburg* disaster, recalled Marshall Göring's visit to the airship hangars on 1 March 1940. Despite the captain's pleas that Göring at least "let the enemy destroy the ships,"[171] the air minister would not change his mind. His primary objection to the airship was its lack of military potential, although like Hitler, he was also afraid of it.[172] Concluding his visit, he summed up the Nazis' real attitude toward the airship by pointing to a Messerschmitt 109 fighter and saying, "This thing is much better."[173] On 6 May 1940, the third anniversary of the *Hindenburg* crash, Wehrmacht demolition specialists blew up the airship installations in Frankfurt, ending the rigid-airship era. There was no public outcry. The French campaign was in the making, and most of the airshipmen and mechanics who might have protested were either mobilized or working in war production.

A symbol that had in sense arrived with a bang, that of Zeppelin LZ 4 crashing at Echterdingen, exited with a whimper in May 1940. Ironically, that same year the technology historian Conrad Matschoss lauded the efforts of German industry and of smaller museums for striving to preserve technological monuments.[174] All that remained in the case of airships was a series of vivid images in the collective memory.

Despite the evolution of the Zeppelin symbol into a direct instrument of political power, popular perception placed it closer to a sign of better times than of domination. Just as ZR 3's flight to America in 1924 or the *Graf Zeppelin*'s visit to the Rhineland in 1930 had served as a positive metaphor, so the newly constructed LZ 129 *Hindenburg* was greeted as a "link through the skies between nations and peoples" that would "solicit [goodwill] for our Germany."[175] Precisely for that reason, the machine fit into the Nazi scheme of things less well than its creator, Count Zeppelin, did. While the Nazi rhetoric stressed that technology was part of German blood and soil, a position that matched the outlook of such figures as the writer Ernst Jünger, Count Zeppelin and his machines did not fit this view, despite attempts to place the count in the Nazi pantheon. He had succeeded much earlier in incorporating his invention into a different German tradition. Long before the Nazis came to power the airship had served as a liberator of sorts in bringing a feeling of joyful fascination to most who saw it. It embodied modernity despite its speed handicap, which would worsen as the airplane evolved. Yet the airship's capacity to stir the imagination should not be discounted. Technology, in this case the Zeppelin, was revolutionary, but politics was reactionary.[176]

Even though the airship had long had a place in the German cultural psyche, the new thinking could not let it remain there on its own. "Technology in itself does not exist; alone, it is fantasy. Every technological idea finds its accomplishment only in reality," wrote one Nazi party official.[177] The *Hindenburg* was the ultimate embodiment of airship technology, which would likely have lasted another decade before the airplane took over. The airship's cost, combined with its limited speed and lack of obvious military potential, made it unattractive to the Nazi regime. While the dirigible could find a role in the realm of Nazi economics, its symbolic value remained fleeting and difficult to control: a memory of the Wilhelminian and Weimar eras, it linked the generations through good and bad periods.

Mircea Eliade reminds us that while images can be manipulated, they can also keep cultures open to new ideas.[178] The dirigible did not just encourage escapism in the worst of times, it also inspired new business and technological ideas in the best of times. Modernist as it was, the Zeppelin symbolized technology in the service of an enlightened, albeit patriotic, world long before assertions about the threat of nationalism were made. Its misuse does, however, confirm that modernity was not "a one-way trip to freedom."[179] Indeed, in the case of the Zeppelin, the Nazis initially supported the airship idea, but only to differentiate themselves from "degenerate" Weimar. Their other actions, though, eventually destroyed the airship as surely as the accident at Lakehurst did. Modernization, then, was a smoke screen rather than a concrete effort. In the case of the airship, the Nazis did not build associations between technology and culture but rather drove them apart. The Nazi version of nationalist resurgence had a use for the airship, but its established message of peace and business was superfluous.

Conclusion

I T IS TEMPTING, IN HINDSIGHT, TO DECLARE THAT THE dirigible in any form was doomed to failure. Yet until the loss of the *Hindenburg* in 1937, German and other advocates of lighter-than-air flight had reason to remain steadfast in its support. Not all its adherents were giddy victims of Zeppelin fever; we must remember that the airship's development was neither simple, linear, nor foreordained, one way or another, and that any explanation for the ultimate supremacy of the airplane must include—besides the technological—social, economic, political, and cultural considerations. Although technologically the Zeppelin led to a dead end (we may think of it as a technological anomaly), these giant craft reflected both the self-interpretation of early-twentieth-century Germans and the hopes of a society wrestling with the awesome challenges of modernity.[1] The context was the early age of aircraft, when the potential of aeronautics remained unexplored and when seeing a flying machine, let alone boarding one, caused an extraordinary excitement.

Indeed, the Zeppelin offers an interesting case study of how society and technology play upon one another. As a scholarly approach this focus on interplay has become known as the social construction of technology.[2] Deriving from recent work in American history of technology and European sociology of scientific knowledge,[3] social construction views technology on three levels: artifact, social processes, and "know-how" (how to apply the artifact to its environment). By this multilayered approach, technological objects are open to analysis not only in their design but also in their uses: how they bridge differences within society, and how consumers of the machine (as well as its engineers) define its value and application.[4]

Yet the analysis must surely extend beyond defined actors like manufacturers and customers; we would paint an incomplete picture of the Zeppelin if we ignored those people who had no direct relationship to the artifact. For this reason, we must incorporate elements from the history of everyday life, which allows the historian to focus on popular reactions and thus come to an understanding of an artifact's or machine's multiple meanings.[5] Through a combination of public events and consumerism, the Zeppelin truly became part of German popular culture, and the Zeppelin sublime surfaced in many forms. The Zeppelin carried heavy symbolic weight, and as a symbol it had the effect of setting and keeping ideas in motion. In this way, the dirigible, like all artifacts, acquired new meanings, some expected and some unintended.

Those meanings varied over time, depending at first on the public character of Count Zeppelin himself. While officialdom in his day associated the airship with national goals, as was the inventor's wish, popular patriotic support derived from the cult surrounding Count Zeppelin. The association of the man and his machine remained powerful—as also happened later and elsewhere with Charles Lindbergh and his *Spirit of St. Louis*. Lindbergh's abilities alone would not have sufficed to captivate the public. His solo crossing of the Atlantic in 1927 needed to coincide with historical circumstances through which the public, captivated by the nature of the events, assigned a longer-lasting symbolic value to the hero, his airplane, and his actions.[6] As stories of "Lucky Lindy" reveal, Lindbergh became a metaphor for various tensions existing in American society that led to a questioning and reexamination of the nation's path.

Newly unified Germany, struggling to define itself, provided a similar backdrop for a man and his machine. Like Lindbergh, Count Zeppelin remained generally committed to the values that had shaped him while at the same time dedicating himself to the pursuit of technology. Unlike Lindbergh, however, Zeppelin resolved rather than raised many of these issues. His origins, experience, and behavior helped to transform him into a myth; indeed, his name displaced that of all other airship constructors, eventually becoming a common noun. Popular culture, although not solely responsible for turning the count into an icon, contributed substantially to this process.

That the count was quickly fashioned into a hero after 1908 is well recognized, but less well understood are the dynamics of his transformation. As the Zeppelin cult took hold, German citizens began to invest in him defining features of German national identity: self-sacrifice, devotion to work and duty, and loyalty to the Fatherland. The rise of mass consumerism and popular culture—particularly in the form of contemporary kitsch and local sports associations founded "in the tradition of Count Zeppelin"—worked to reinforce the count's appeal. To be German was to revere Count Zeppelin. One demonstrated attachment to the new German nation through devotion to him (but also to his machines). After Zeppelin's death, his successor, Hugo Eckener, capitalized on the count's image to gain public support for civilian airship links. Later, National Socialists sought to use the count's military past as a brick in the construction of the Nazi pantheon.

In parallel to Count Zeppelin's image, the airship in its various models also acquired multiple meanings, further reflecting German interpretations of a modern world. From machines of escapism and showpieces at patriotic festivals, dirigibles were transformed into "superweapons" immediately before World War I. Portrayed alternately as a troop carrier, a heavy cruiser, or a light scout-ship, the Zeppelin reinforced the idea of German military invincibility and helped prepare the country for war. Although expectations about the airship's role stood in stark contrast to military realities, as a psychological weapon the dirigible did not disappoint. After 1918, Allied anger was such that Germany was barred from building airships for its own uses until 1926. Two years before the end of the ban, public enthusiasm rose when the United States took delivery of a reparations Zeppelin. When channeled into fund-raising efforts and the promise of new adventures in the form of Arctic exploration, such devotion turned the machine into a kind of national remedy for lost war.[7] Although financially troubled, the Weimar government nonetheless allocated funds to the design and construction of a "transport prototype," the soon-to-be world-famous LZ 127 *Graf Zeppelin*.

While Zeppelins served as nationalist icons from 1908 onward, under the Nazis their political use became blatant and conspicuous. The airship offered a seemingly perfect propaganda piece for National Socialist leaders, appearing at Nazi rallies, whipping up nationalist enthusiasm before the arrival of Hitler.

Even so, the Zeppelin was a problematic symbol of National Socialism, and a number of factors complicated its use as a tool of the Third Reich. Ironically, the *Hindenburg* disaster in the long run served Nazi designs well, providing the Nazis a pretext for abandoning the airship.

The end of the airship era did not mean the end of the Zeppelin company. During World War II the firm and its subsidiaries produced parts for V-2 rockets and aircraft. Allied bombings eventually leveled the factory, dashing any hopes Eckener may have had of restarting a civilian plant after the war. The untiring businessman instead served briefly as a consultant to the Goodyear corporation in the United States. As nominal head of the Zeppelin works, he was eventually convicted of helping the war effort. Although such postwar politicians as Theodor Heuss, the first president of the Federal Republic, and Konrad Adenauer, its first chancellor, intervened in his behalf, Eckener lost his civil privileges in 1948 for five years and had to pay a 100,000-mark fine.

Eckener, like other major figures of the Zeppelin era, was past retirement age when the Federal Republic of Germany was founded in 1949, and he removed himself from industrial activities. He and his former business rival Alfred Colsman became active in local political circles.[8] Along with the company's chief engineer, Ludwig Dürr, they accepted occasional honors for their past airship activities.[9] All three died quietly in the early 1950s.

"The [airship] episode was interesting, but it has now become history (almost legend)," wrote an aging Eckener to a former colleague in 1950.[10] As if in response, various new associations intent on perpetuating enthusiasm for Zeppelins were founded in Germany and the United States at about this time and began meeting at regular intervals, attracting airship veterans as well as younger enthusiasts. They shared a "what if" sense of nostalgia.[11] For almost two decades after the war, many of them searched for an opportunity to resurrect the airship yet again. According to rumor and claims that enthusiasts made in private correspondence, fantastic new dirigibles would soon replace seagoing cargo vessels and even use atomic-powered engines.[12]

Considering the linked themes of the popular perception of a machine, its political uses, and its engineering reality, this study demonstrates how a machine that was neither practical nor designed for mass consumption nevertheless assumed a central place in determining European understandings of aviation technology. For all its peculiarities, such a case offers a model that can be transposed to other technology projects.

During the first Zeppelin era, ocean liners came to represent the greatness of Western nations in peacetime. Sophisticated engineering affirmed dominance over nature, claiming such impossible feats as making a ship unsinkable. This very notion, contradicted through the tragedy of the SS *Titanic* in 1912, made the ship more famous than if it had steered clear of the iceberg. Indeed, its sinking prompted discussions about the arrogance of riches and the breakdown of moral and social codes in society as a whole.[13] Later accounts and films showed that even accounting for manipulation of consumers, popular culture fed willingly (and morbidly) on representations of the event. Yet the romance of sea cruises did not disappear. Rather, the tragedy became a reference point for matters of technical failure as well as their social dimensions.

Seventy-four years after the *Titanic* went down, a failure in the American space program again underscored the importance of great technological symbols. Whereas the massive *Saturn V* moon rocket hardly earned a place in popular culture (despite its having paved the way for the stupendous moon-landing achievement), the space shuttle achieved iconic status. Sold to the public as a machine that would drive one into orbit for breakfast and home for dinner, it fell far short of expectations from the very start of operations in 1981. However, when, in 1986, the shuttle *Challenger* disaster exposed a series of serious technical and management failures, public shock gave way to unwavering support for the program. To argue that such public approval resulted from governmental and media manipulation would be both cynical and inaccurate. One must account for other factors, which range from the presence of a teacher among the martyred astronauts, to the association of the machine with patriotic reawakening following the post–Vietnam War melancholy. A technological tragedy thus adds to the sublime effect of a machine by turning it into a human metaphor: we humanize flawed machines, and in so doing reinforce their symbolic power.

Other examples exposing the importance of a cultural public dimension in technology and its sublime dimension take us back to Europe. In 1998 the derailment of a high-speed German ICE train killed one hundred passengers and destroyed the myth of ultramodernity associated with the new rail system. In the age of machines, accidents are unavoidable, but ICE had carved a special niche in German popular culture, where trains have often held a special place. Where ICE went, modernity reigned; where it did not, backwardness endured. The accident did not disrupt high-speed train travel, but it modified public fascination by reawakening an element of doubt and ending a dream of clean, always benevolent technology: machinery remains only as good as its designers make it.[14]

Two years after the ICE catastrophe, thousands of witnesses watched as a burning Concorde supersonic passenger jet plowed into the ground shortly after takeoff from Paris. The accident was but another aviation tragedy, yet it made headlines because this first crash of a Concorde also destroyed another myth of indestructibility. A Franco-British project from the 1960s designed in part as a response to American dominance of the civil-aviation market, the machine was a commercial failure (only the national French and British carriers bought it). Yet it acquired a special identity in both countries. Founded on the elements of rarity (only twenty Concordes were built), elitism (few could afford to fly it), security, and speed, this symbol of technological modernity strangely echoed the appeal of the great Zeppelins seventy years earlier. The Paris crash did not just kill people, then; it crushed another dream of flight that reflected the popular pride associated with the sublime (and rather noisy) effect of these machines in operation.[15]

Technological failure is not a requirement for understanding the appeal of certain machines, but it does help one gauge the level of enthusiasm, just as does success. The technological sublime of a space launch or the myth of the astronaut as frontier pioneer recalls the place the Zeppelin and its crew maintained in German hearts. Cultural, nontechnical questions can matter as much as design issues in accounting for the success or failure of a technology in the public eye.

The philosopher Georges Perec once marveled at people's capacity for developing an affection for *things*. In the case of the airship, the love affair continued well beyond the end of the golden age of dirigibles. Its sublime effect carried on despite its absence from the skies. Zeppelin's name is a common noun in many languages. In Vienna, bakeries offer "Zeppelin" loaves, and in Frankfurt, Stephan Weiss's "Zeppelin Wurst," first introduced in 1909, is still available throughout the city. Artists have adopted the airship as a source of inspiration, and writers such as Nobel Prize winner Günther Grass associate it readily with German identity.[16] Postal services issue commemorative stamps depicting the rigid airship, while historians and engineers continue to marvel at either the potential or the uselessness of the machine's design. Late in his life, Eckener wondered whether people would go beyond the sublime and re-member both the technical and the moral and political aspects of the airship.

He need not have worried. At the end of the twentieth century, as new mu-seums and exhibits devoted to airships opened their doors in Germany,[17] one did not have to search hard for evidence that the Zeppelin continued to stir the imagination.

ABBREVIATIONS

AA   Auswärtiges Amt (German Ministry of Foreign Affairs)

ACD   Archiv für Christliche Demokratische Politik (Archives of Christian Democratic Politics), St. Augustin, Germany

AdR   Archiv der Republik (Archives of the Republic), Vienna

ASTRA   Air and Space Transport Research Archive, Geneva

BA   Bundesarchiv (German Federal Archives), Koblenz

BA/MA   Bundesarchiv/Militärarchiv (German Military Archives), Freiburg im Breisgau

BDM   Bibliothek des Deutschen Museums (Deutsches Museum Library), Munich

DGLR   Deutsche Gesellschaft für Luft- und Raumfahrt (German Air and Space Society), Bonn

DLH   Deutsche Lufthansa Archiv (Deutsche Lufthansa Corporate Archives), Cologne

DNB   Deutsches Nachrichten Büro (German Press Agency), 1933–45

DZR   Deutsche Zeppelin Reederei (German Zeppelin Line)

ETH   Eidgenössische technische Hochschule (Federal Institute of Technology Archives), Zürich

GStA   Geheimes Staatsarchiv (State Archives of Prussia), Merseburg

HAStK   Hauptarchiv der Stadt Köln (Central Archives of the City of Cologne)

HHW   Hessisches Hauptstaatsarchiv (Hessen Central State Archives), Wiesbaden

HUH   Harvard University, Houghton Library, Cambridge, Mass.

KA   Kriegsarchiv (Austrian War Archives), Vienna

LCM   Library of Congress, Manuscripts Division, Washington, D.C.

LHK   Landeshauptarchiv (Central Regional Archives), Koblenz

LZA   Luftschiffbau Zeppelin Archiv (Zeppelin Museum Archives), Friedrichshafen

MHS   Maryland Historical Society, Baltimore

NARA   National Archives and Records Administration, College Park, Md.

NASM   National Air and Space Museum, Washington, D.C.

OSA   Österreichisches Staatsarchiv (Austrian State Archives), Vienna

PA   Politisches Archiv (German Foreign Ministry Archives), Bonn

PRO   Public Records Office, Kew

RFM   Reichsfinanzministerium (German Ministry of Finance)

RLM   Reichsluftfahrtministerium (German Air Ministry), 1933–45

RMI   Reichsministerium des Innern (German Interior Ministry)

RMVP   Reichsministerium für Volksauklärung und Propaganda (German Ministry of Propaganda), 1933–45

RPM   Reichspostministerium (German Postal Ministry)

RVM   Reichsverkehrsministerium (German Transportation Ministry), 1919–33

RWM   Reichswehrministerium (German Defense Ministry), 1919–33

SFA   Swiss Federal Archives, Bern

SHAA   Service historique de l'Armée de l'air (Archives of the French Air Force), Château de Vincennes, Paris

SML   Science Museum Archives, London

SWA   Stiftung Westfälisches Wirtschaftsarchiv (Westphalian Business Archives Foundation), Dortmund

SWB   Stadtarchiv und Wissenschaftliche Bibliothek (Bonn City Archives), Bonn

TMW   Technisches Museum, Wien (Technology Museum, Vienna)

UTD   University of Texas, Dallas

VDI   Verein Deutsche Ingenieure (Association of German Engineers)

WSU   Wright State University Library Archives, Dayton, Ohio

WTB   Wolffs Telegraphisches Büro (Wolff Press Agency), 1919–33

NOTES

Introduction: Visions of the Sublime

1. Hugo Eckener, *Im Zeppelin über Länder und Meere* (Flensburg: Christian Wolff, 1949), 108–9, 112–13.

2. Lewis Mumford, *Technics and Civilization* (London: Routledge, 1947), 45.

3. David Nye, *American Technological Sublime* (Cambridge: MIT Press, 1994), xiii.

4. Ibid.

5. Gert Mattenklott, "On the Ideology of Bigness," *Daedalus* 61 (1996): 78–79.

6. Nye, 42–43, 45ff., 77ff., 225ff.

7. Bernd Nicolai, "National Bridgings: Technology and Social Identity in the Bridge Building of the Nineteenth Century," *Daidalos* 57 (September 1995): 94–103.

8. Eckener, 112–13.

9. Mumford, introduction.

10. See the "Essay on Sources and Methods" at the end of this book.

Chapter One. Balloons into Dirigibles

1. William C. Carter, *The Proustian Quest* (New York: New York University Press, 1992), 2, 7.

2. Roger Cooter and Stephen Pumfrey, "Separate Spheres and Public Places: Reflections on the History of Science Popularization and Science in Popular Culture," *History of Science* 32 (1994): 249–50.

3. Emmanuel Chadeau, *Le rêve et la puissance: L'avion et son siècle* (Paris: Fayard, 1996), 22–23.

4. John Anderson, *A History of Aerodynamics and Its Impact on Flying Machines* (New York: Cambridge University Press, 1998), 114, 138.

5. Donald J. Bush, *The Streamlined Decade* (New York: Braziller, 1975), 6.

6. Winfried de Fonvielle, *Histoire de la navigation aérienne* (Paris: Hachette, 1907), 24–26.

7. Hermann Moedebeck, *Die Luftschifffahrt in ihrer neuesten Entwickelung* (Berlin: Mittler, 1887), 34–35.

8. E. J. Marey, *La machine animale* (Paris: Alcan, 1886). Octave Chanute, the flight theoretician credited with furnishing several solutions to the Wright brothers, also concentrated on the flight of gulls, sending photographs to several of his correspondents with comments attached.

9. Gerhard Wissmann, *Geschichte der Luftfahrt,* 2d ed. (East Berlin: VEB, 1966), 129–37; Douglas Robinson, *Giants in the Sky: A History of the Rigid Airship* (Seattle: University of Washington Press, 1973), 1–8; Günter Schmitt, "Vom Ballon zum Luftschiff," in *Leichter als Luft-Ballone und Luftschiffe,* ed. Dorothea Haaland et al. (Bonn: Bernard & Graefe, 1997), 74–85; Jules Duhem, *Histoire de l'arme aérienne avant le moteur* (Paris: Nouvelles éditions latines, 1964), 107–15, 278–83.

10. Jennifer Tucker, "Voyages of Discovery on Oceans of Air: Scientific Discovery and the Image of Science in an Age of 'Balloonacy,'" *Osiris* 11 (1996): 144–76; BDM, 1910 B98, "Patentschriften betreffend Luftschiffahrt."

11. Quoted in Moedebeck, 1.

12. Henry Dale, *Early Flying Machines* (New York: Oxford University Press, 1992), 25.

13. Wolfgang Schivelbusch, *The Railway Journey* (Berkeley: University of California Press, 1986), 9, 33; M. J. B. Davy, *Interpretive History of Flight* (London: HMSO, 1937), 56–57.

14. Wissmann, 138–39; Gero von Langsdorff, "Angebote auch an die Regierung Preussens: Neue Funde zu den Luftschiffprojekten des Ingenieurs Paul Hänlein," in *Lenkballone vor Zeppelin,* ed. Helmut Schubert (Bonn: DGLR, 1992), 10.

15. Serge Picard, "Le moteur Lenoir," *La revue* 16 (September 1996): 45–53.

16. TMW, GSA 241, *Neues Wiener Tagblatt,* 5 and 6 November 1908; Wilhelm Hänlein, "Das lenkbare Luftschiff von Paul Hänlein," *Flug- und Motor-Technik,* 25 November 1909, 1–8.

17. LCM, Tissandier papers, box 29; Gaston Tissandier, *Histoire de mes ascensions* (Paris: Dreyfous, 1888), 232–68; Rodolphe Soreau, *Le problème de la direction des ballons* (Paris: Librairie centrale des sciences, 1893), 20–21.

18. Soreau, 22.

19. Heinz Gartmann, *Science as History* (London: Hodder & Stoughton, 1961), 99–100; Anderson, 100.

20. Langsdorff, 10–12; Leo Koenigsberger, *Hermann von Helmholtz* (1906; reprint, New York: Dover, 1965), 296–98.

21. James Martin Hunn, "The Balloon Craze in France, 1783–1799: A Study in Popular Science" (Ph.D. diss., Vanderbilt University, 1982).

22. Helmut Reinicke, *Aufstieg und Revolution* (Berlin: Transit, 1988); cf. Jürgen Link, "Einfluss des Fliegens!—Auf den Stil selbst!" and Hans Ulrich Seeber, "Der Ballonaufstieg als Spektakel und Metapher," in *Bewegung und Stillstand in Meta-*

phern und Mythen, ed. Jürgen Link and Wulf Wülfing (Stuttgart: Klett-Cotta, 1984), 149–63 and 164–200.

23. Friedrich Nietzsche, *Morgenröte* (1881), reprinted in *Reisen im Luftmeer,* ed. Karl Riha (Munich: Carl Hanser, 1983), 243.

24. Léo Dex, "Ballons dirigeables et appareils volants," *Magasin pittoresque,* 1896, 312.

25. Merritt Roe Smith, "Technological Determinism in American Culture," in *Does Technology Drive History?* ed. Merritt Roe Smith and Leo Marx (Cambridge: MIT Press, 1994), 8.

26. Judith Shklar, "The Political Theory of Utopia from Melancholy to Nostalgia," in *Utopias and Utopian Thought,* ed. Frank E. Manuel (Boston: Beacon Press, 1965), 101–16; cf. W. H. G. Armitage, "Utopias: The Technological and Educational Dimensions," in *Utopias,* ed. Peter Alexander and Roger Gill (London: Duckworth, 1984), 86.

27. Rolf Peter Sieferle, *Fortschrittsfeinde?* (Munich: Beck, 1984), 143–54.

28. Isaac Asimov, *Future Days: A Nineteenth Century Vision of the Year 2000* (New York: Henry Holt, 1986).

29. Albert Robida, *Le vingtième siècle* (1883; reprint, Paris: Tallandier, 1991); *La guerre au vingtième siècle* (1887; reprint, Paris: Tallandier, 1991); Edward Tenner, "Gezeichnete Zukunft," *Kultur und Technik,* March 1991, 23–29; Daniel Compère, "Les monstres nouveaux," *Romantisme* 13, 41 (1983): 91–100.

30. I. F. Clarke, ed., *The Tale of the Next Great War, 1871–1914* (Syracuse, N.Y.: Syracuse University Press, 1995), 1–2; H. Bruce Franklin, *War Stars: The Superweapon and the American Imagination* (New York: Oxford University Press, 1988), 19–53.

31. Michael Paris, *Winged Warfare: The Literature and Theory of Aerial Warfare in Britain, 1859–1917* (Manchester: Manchester University Press, 1992), 19.

32. W. Behringer and C. Ott-Koptschalijski, *Der Traum vom Fliegen zwischen Mythos und Technik* (Frankfurt a.M: Fischer, 1991), 355–56; Edgar Allan Poe, *The Science Fiction of Edgar Allan Poe* (New York: Penguin, 1976).

33. Laurence Goldstein, *The Flying Machine and Modern Literature* (Bloomington: Indiana University Press, 1986), 55, 67, 69; Anthony Frewin, *One Hundred Years of Science Fiction Illustration* (New York: Pyramid, 1974), 45; Jules Verne, *Paris au XXe siècle* (Paris: Hachette/Le cherche midi, 1995). This unfinished novel includes a depiction of weather balloons that guard a city from lightning strikes but by their very presence send its inhabitants into a funk.

34. Carl Falkenhorst, *Luftfahrten* (1891; reprint, Düsseldorf: VDI, 1987), 130–37.

35. Ibid.

36. Paris, 125–26.

37. Ibid., 26–27.

38. George Wise, "Technological Prediction, 1890–1940" (Ph.D. diss., Boston University, 1976), 115–16.

39. A. Hildebrandt, *Luftschiffahrt* (Munich: Oldenbourg, 1909), 345–403.

40. Bernard Korzus, ed., *Leichter als Luft: Zur Geschichte der Ballonfahrt* (Münster: Westfälisches Landesmuseum für Kunst und Kulturgeschichte, 1985), 105–35; Paul Wider, *Menschen und Ballone* (Munich: Bechtle, 1993), 119–23.

41. *Die Gartenlaube* 37 (1862): 568–69; 38 (1862): 585–88; R. Güring, "Wissenschaftliche Ballonhochfahrten," in *Wir Luftschiffer,* ed. Karl Bröckelmann (Berlin: Ullstein, 1909), 48–65.

42. Wider, 171–76.

43. Jean Louis Schlimm, *Ludwigs Traum vom Fliegen* (Oberhaching: Aviatic Verlag, 1995), 54–74.

44. BDM, *Münchner Verein für Luftschiffahrt Jahresbericht, 1899–1900* (Munich: Verein für Luftschiffahrt, 1900). The association began in 1889 with approximately two hundred members. In 1900, Count Zeppelin was member 398 out of 401.

45. Hans G. Knäusel, *Zeppelin and the United States of America: An Important Episode in German-American Relations,* 2d ed. (Friedrichshafen: Zeppelin, 1981), 15; Rhoda R. Gilman, "Count Zeppelin and the American Atmosphere," *Smithsonian Journal of History* 3, 1 (Spring 1968): 29–40; Lutz Tittel, *Graf Zeppelin: His Life and His Work,* trans. Peter A. Schmidt (Friedrichshafen: Zeppelin Museum, 1995), 17–18.

46. Heinrich von Stephan, *Weltpost und Luftschiffahrt* (Berlin: Springer, 1874), 1, 51–53, 74–75.

47. Hans G. Knäusel, "Der Zeppelin-Luftschiffbau im Spiegel der Fachliteratur," in Haaland et al., 277–78; cf. Tittel, *Graf Zeppelin,* 25; Rolf Italiaander, *Ferdinand Graf von Zeppelin* (Konstanz: Stadler, 1986), 62–63.

48. Robinson, 14; Henry Cord Meyer, *Airshipmen, Businessmen, and Politics, 1890–1940* (Washington, D.C.: Smithsonian, 1991), 31.

49. Hans G. Knäusel, *LZ 1, der erste Zeppelin: Geschichte einer Idee, 1874–1908* (Bonn: Kirschbaum, 1985), 43–45; Meyer, 27–30.

50. NASM, A3G-309030-01, "To Echterdingen," published for Count Zeppelin's seventy-fifth birthday. The main exception to the trend of the mythical accounts is Knäusel, *LZ 1.*

51. "Patentschrift No 98580 Klasse 77 Sport, Graf v. Zeppelin in Stuttgart," reprinted in Knäusel, *LZ 1,* 50–53.

52. Robinson, 21.

53. Friedrich Sass, *Geschichte des deutschen Verbrennungsmotorenbaues von 1860 bis 1918* (Berlin: Springer, 1962), 363.

54. Ibid., 11ff., 245ff.

55. NASM, A3G-309040-01, Office of the U.S. Naval Attaché in Berlin, trans-

lation of Müller-Breslau, "On the History of the Zeppelin Air Ship," 1924; *Zeitschrift des VDI*, 1908, 1549.

56. Letter of Count Zeppelin to the War Ministry, 15 August 1895, reprinted in *Die Militärluftfahrt bis zum Beginn des Weltkrieges* (Frankfurt: Mittler, 1966), 31–33.

57. Count Zeppelin, "Entwürfe für lenkbare Luftschiffe," 6 February 1896, reprinted in Knäusel, *LZ 1*, 56–58.

58. Knäusel, "Der Zeppelin-Luftschiffbau," 279–80. Count Zeppelin's August 1895 patent 98580 bore a close resemblance to patent 91887, granted to Caesar Eggert in Berlin six months earlier.

59. *VDI Aufruf*, 30 December 1896, reprinted in Knäusel, *LZ 1*, 63.

60. Claus Laske et al., *Ernst Georg Baumgarten—Ein sächsischer Flugpionier* (Karl-Marx-Stadt: n.p., n.d.).

61. Knäusel, *LZ 1*, 33–40; cf. Cvi Rotem, *David Schwarz, die Tragödie des Erfinders* (Bloomington, Ind.: n.p., 1983); G. Wahl, "Zur Geschichte des Leinbergerschen Luftschiffes und zur Geschichte des Metallballons überhaupt," *Archiv für die Geschichte der Naturwissenschaften und der Technik* 2, 5 (July 1910): 387–93.

62. Schmitt, 106–8; cf. Wissmann, 143–44. Other more sophisticated aluminum airship designs existed, including those of the aerospace pioneer Constantin Tsiolkowski; however, Tsiolkowski failed to obtain financial support from Russian authorities.

63. Gustav von Falkenberg, *Elektrizität und Luftschiffahrt in ihren wechselseitigen Beziehungen* (Rostock i.M.: Volckmann/Wette, 1910), 37.

64. Knäusel, *LZ 1*, 64–76.

65. Ibid., 77–87. The "Gesellschaft zur Förderung der Luftschiffahrt" had a capital base of 22,000 marks.

66. Helmut Maier, "Aluminum Looks for a Use: Symbolism or Calculation?" paper read at the Society for the History of Technology annual meeting, 8 October 1994.

67. M. J. French, *Invention and Evolution* (New York: Cambridge University Press, 1988), 54–67.

68. Knäusel, *LZ 1*, 99–104.

69. NASM, A3G-309201-01, Gustav Levering, "The Ship That Flies," *Pearson's Magazine*, November 1900; *Die Umschau*, 9 August 1900, 644–48.

70. *Die Gartenlaube*, supplement, 16 (1900): 518.

71. *Die Umschau*, 9 August 1900, 644–48; Hermann Hoernes, *Lenkbare Ballons* (Leipzig: Engelmann, 1902); Herrman Moedebeck, *Die Luftschiffahrt* (Strasbourg: Trübner, 1906); TMW, Kress papers, *Zeitschrift für Luftschiffahrt* 19 (1900): 228–29; Viktor Silberer, *Der heutige Stand der Luftschiffahrt* (Vienna: Allgemeine Sport Zeitung, 1900, 1904, 1905). Vocal opponents included Austrian captain Franz Hinterstoisser of the military-aeronautical section and aeronaut Viktor Silberer.

72. TMW, Kress papers, Paul Pacher, "Graf Zeppelin und seine Neider," *Ostdeutsche Rundschau,* 1 August 1900.

73. LZA, 04/75, 1892 school notebook cover recounting Count Zeppelin's actions at Schirlenhoff (25 July 1870).

74. The Wright brothers' experiments in Dayton, Ohio, in 1904–5 also failed to generate much public interest. Joseph Corn, *The Winged Gospel: America's Romance with Aviation, 1900–1950* (New York: Oxford University Press, 1983), 7.

75. Roger Chickering, *We Men Who Feel Most German: A Cultural Study of the Pan-German League, 1886–1914* (Boston: Allen & Unwin, 1984), 16.

76. Lewis Mumford, *Technics and Civilization* (New York: Harcourt & Brace, 1963), 336–37; Sieferle, 157.

77. David Blackbourne and Geoff Eley, *The Peculiarities of German History* (New York: Oxford University Press, 1984), 185–87.

78. Lawrence W. Levine, "The Folklore of Industrial Society: Popular Culture and Its Audiences," *American Historical Review* 97, 5 (December 1992): 1375.

79. Charles H. Gibbs-Smith, *The Rebirth of European Aviation, 1902–1908* (London: Science Museum, 1974), 7–12.

80. A. Stolberg, "Die letzten Aufstiege des Zeppelinschen Luftschiffes," *Die Umschau,* 1 December 1900, 966.

81. Count Zeppelin, *Ueber die Ansicht auf Verwirklichung und den Wert der Flugschiffahrt* (Berlin: Deutsche Kolonialgesellschaft, 1901), reprinted in Knäusel, *LZ 1,* 157–62.

82. HUH, Abott Lawrence Rotch papers, 1903 correspondence with Count Zeppelin. Zeppelin even attempted to obtain foreign funding, but this led nowhere. Rotch, a pioneer meteorologist, looked into the possibility of getting American funding for the count to exhibit his invention and raise funds at the St. Louis World Exposition in 1904.

83. Knäusel, *LZ 1,* 189–90.

84. LZA, 04/174, *Börsen und Handelsblatt,* 22 September 1903.

85. LZA, 04/118, letter of Scherl to Count Zeppelin, 13 June and 10 August 1903; cf. Wolfgang Meighörner, " . . . *der Welt die Wundergabe der Beherrschung des Luftmeeres schenken": Die Geschichte des Luftschiffs LZ 2* (Friedrichshafen: Zeppelin Museum, 1991), 18.

86. LZA, 04/118, letter of Scherl to Count Zeppelin, 15 August 1903; cf. Karl Clausberg, *Zeppelin: Die Geschichte eines unwahrscheinlichen Erfolges* (1979; reprint, Augsburg: Weltbild, 1989), 37.

87. Meighörner, 15–17.

88. W. Rickmers, *Die Beherrschung der Luft* (Vienna: Eduard Beyer, 1903), 1.

89. Meighörner, 61–77; Meyer, 33.

90. August von Parseval, "Erfahrungen mit dem Parseval auf der 'Ila,'" *Flug- und Motor-Technik,* 10 November 1909, 14–15.

91. L. Sazerac de Forge, *La conquête de l'air* (Paris: Berger-Levrault, 1907), 191–92.

92. Memorandum of Chief of Staff von Schlieffen, 1 November 1905, and evaluation of Zeppelin concept by Major Gross, 8 December 1905, reprinted in *Die Militärluftfahrt*, 33–37.

93. Meighörner, 41.

94. The mistake is partly due to the fact that few pictures of LZ 2 exist, and that the successor ship LZ 3, which incorporated the change, had similar dimensions. Incorrect illustrations and descriptions appear in most English- and German-language technical histories. Exceptions include Robinson and Meighörner.

95. Meighörner, 23.

96. Peter Kleinheins, ed., *Die Grossen Zeppeline* (Düsseldorf: VDI, 1985), 24–26.

97. BA/MA, PH9–V/48, minutes of Airship Commission meeting, 24 February 1906.

98. Ansbert Vorreiter, "Konstruktionsprinzipien der Motoren für Luftschiffe und Flugapparate," *Flug- und Motor-Technik,* 10 October 1909, 14.

99. Knäusel, *LZ 1,* 198–200; Jürgen Eichler, *Luftschiffe und Luftschiffahrt* (Berlin: Brandenburgisches Verlagshaus, 1993), 37–40. A general survey of German airship types appears in Schmitt, 120–44.

100. Eichler, 40; Gibbs-Smith, 199–232; Sazerac de Forge, 191–92. In the society's name (Motor Luftschiffahrt Studien Gesellschaft), the term *Luftschiffahrt* was used to refer to all flying machines, not just airships.

101. Eichler, 43–44; Meyer, 34; Clausberg, 42.

102. BA, R5/3829, *Neue Zürcher Zeitung,* 12 October 1906. An open letter to the emperor included the endorsements of scientists such as Ludwig Prandtl, Fritz Klein, and Hugo Hergesell.

103. Meighörner, 80.

104. "Lenkbare Luftballons im Jahre 1903," *Die Umschau,* 19 December 1903, 1031–35; Guido Castagneris, "On the Conditions of Equality of Statistical Stability, between the Dirigibles 'Patrie' and 'Zeppelin,'" *Aëronautical Journal,* July 1908, 94.

105. Ludwig Dürr, "Fünfundzwanzig Jahre Zeppelin-Luftschiffbau," in Kleinheins, 73; *Die Umschau,* 5 October 1907, 835–37.

106. Robinson, 34.

107. BA, R5/3829, letter of Chancellor von Bülow to the kaiser, 25 February 1907; cf. Eichler, 44–46. The net income from the lottery was 150,000 marks, well short of the 1 million marks the count said he needed to draft and build improved airship designs.

108. *Reichstag Berichte, 92. Sitzung,* 1 February 1908, 2810–12, comments of Reichstag member Hug; BA, R5/3846, RMI protocol, 22 March 1908.

109. BA, R/3830, *Berliner Neueste Nachrichten,* 6 February 1908; cf. *Reichstag Berichte, 121. Sitzung,* 13 March 1908, 3871–73.

110. BA, R5/3839, *Berliner Lokal-Anzeiger,* 24 July 1908.

111. Wolf Volker Weigand, *Walter Wilhelm Goetz, 1867–1958* (Boppard a.R.: Boldt, 1992), 125. Goetz, then a professor in Tübingen, accompanied a university delegation to Friedrichshafen in March 1908 with the intention of reading a poem he had composed for the count.

112. HHW, 405I-P/4, vol. 2496, letter of the Mittelreheinischer Verein für Luftschiffahrt (Mid-Rhine Airship Association), Mainz, 3 June 1908, which notes that membership jumped from thirty-two hundred to five thousand in 1907; Dr. Kuhtz, "Was muss man vom lenkbaren Luftschiff wissen?" *Die Umschau,* 5 October 1907, 809.

113. *Kladderadatsch,* 21 October 1906.

114. Rudolf Martin, *Berlin-Bagdad: Das deutsche Weltreich im Zeitalter der Luftschiffahrt* (Stuttgart: Deutsche Verlagsanstalt, 1907); cf. Clausberg, 133; Robert Wohl, *A Passion for Wings: Aviation and the Western Imagination, 1908–1918* (New Haven: Yale University Press, 1994), 76–81, 89–94.

115. BA, R5/3840, "Englands Weltmacht und die Motorluftschiffahrt," *Nord Süd* 32, 1 (1908): 63–72.

116. Claus Ritter, *Kampf um Utopolis oder die Mobilmachung der Zukunft* (East Berlin: Verlag der Nation, 1987).

117. Martin, 94–95.

118. Stephen Kern, *The Culture of Time and Space, 1880–1918* (Cambridge: Harvard University Press, 1983), 224.

119. Martin, 97.

120. BA, R5/3829, protocol of meeting between Count Zeppelin and Prussian and imperial officials, 19 December 1906; cf. Meyer, 34.

121. Julius Rota, *Die aerodynamische Versuchanstalt in Göttingen, ein Werk Ludwig Prandtls* (Göttingen: Van den Hoeck & Ruprecht, 1990), 18, 35–37.

122. *Tägliche Rundschau* (Berlin), 24 February 1907; cf. BA, R5/3829, *Ueberall: Illustrierte Zeitschrift für Armee und Marine* 9, 17 (19 January 1907).

123. BA, R5/3829, letter of Chancellor von Bülow to the kaiser, 25 February 1907.

124. *Tägliche Rundschau,* 24 February 1907; cf. BA, R5/3829, *Ueberall: Illustrierte Zeitschrift für Armee und Marine* 9, 17 (19 January 1907); *Reichstag Berichte, 21. Sitzung,* 18 March 1907, 568, comments of Freiherr von Richthofen-Damsdorf.

125. *Vossische Zeitung,* 9 August 1908, reviews the various Reichstag decisions to date concerning Count Zeppelin. Among his political supporters were the conservatives Richthofen-Damsdorf and von Gersdorf, the National Liberal von Schubert, and the Social Democrat Singer.

126. BA, R5/3839, clippings collection.

127. BA, R5/3829, protocol of meeting between Count Zeppelin and Prussian and imperial officials, 19 December 1906.

128. Robinson, 35.

129. Tittel, *Graf Zeppelin*, 12–14, 18.

130. Eichler, 42; Lutz Tittel, *Die Fahrten des LZ 4, 1908* (Friedrichshafen: Zeppelin Museum, 1983), 22–28.

131. Tittel, *Die Fahrten des LZ 4*, 29.

132. Alfred Gollin, *No Longer an Island: Britain and the Wright Brothers, 1902–1909* (Stanford, Calif.: Stanford University Press, 1984), chaps. 12 and 13.

133. *Illustriertes Wiener Extrablatt*, 5 August 1908.

134. Peter Fritzsche, *A Nation of Fliers: German Aviation and the Popular Imagination* (Cambridge: Harvard University Press, 1992), 10; Don Handelman, *Models and Mirrors: Towards an Anthropology of Public Events* (1990; reprint, New York: Berghahn Books, 1998), 9.

135. Carl Zuckmayer, *Als wär's ein Stück von mir* (Hamburg: Fischer, 1966), 202–3.

136. Susanne Faschon and Wolfgang Hütte, *Die alte Stadt Moguntia kommt immer mehr zu Ehr* (Mainz: H. Schmidt, 1986), 110–13.

137. Eichler, 49; Tittel, *Die Fahrten des LZ 4*, 37.

138. Bernd Klagholz, *Der Tag von Echterdingen* (Leinfelden-Echterdingen: Stadtarchiv, 1998), 57–59.

139. Tittel, *Die Fahrten des LZ 4*, 38.

140. Klagholz, 74–75.

141. Hugo Eckener, *Graf Zeppelin: Sein Leben nach eigenen Aufzeichnungen und persönlichen Errinerungen* (Stuttgart: Cotta, 1938), 161; cf. Klagholz, 79–83.

142. Fritzsche, 11–16; Clausberg, 47–56; Meyer, 35–37; Robinson, 36–41.

143. Friedrich Rapp, "Humanism and Technology: The Two-Cultures Debate," *Technology in Society* 7 (1985): 429.

Chapter Two. The Machine above the Garden: Airship Culture in Imperial Germany

1. Robert Zedlitz-Trütschler, *Twelve Years at the Imperial German Court* (New York: Doran, 1924), 229.

2. Jack A. Goldstone, *Revolution and Rebellion in the Early Modern World* (Berkeley: University of California Press, 1991), 427; Eric Hobsbawm, "Mass-Producing Traditions: Europe, 1870–1914," in *The Invention of Tradition*, ed. Eric Hobsbawm (New York: Cambridge University Press, 1983), 274.

3. Richard Overy, "Heralds of Modernity: Cars and Planes, from Invention to Necessity," in *Fin de Siècle and Its Legacy*, ed. Mikuláš Teich and Roy Porter (New York: Cambridge University Press, 1990), 74.

4. Helmut Reinicke, "Begeisternde Arbeit und Schwereloser Aufstieg vor 1914," in *Vom Wert der Arbeit: Zur literarischen Konstitution des Wertkomplexes*

'*Arbeit*' in der deutschen Literatur, ed. Harro Segeberg (Tübingen: Niemeyer, 1991), 295.

5. Robert Wohl, *A Passion for Wings: Aviation and the Western Imagination, 1908–1918* (New Haven: Yale University Press, 1994), 54–66; Scott W. Palmer, "On Wings of Courage: Public 'Air-Mindedness' and National Identity in Late Imperial Russia," *Russian Review* 54 (April 1995): 209–26.

6. Peter Fritzsche, *A Nation of Fliers: German Aviation and the Popular Imagination* (Cambridge: Harvard University Press, 1992), 29.

7. Alon Confino, "The Nation as a Local Metaphor: Heimat, National Memory, and the German Empire, 1871–1918," *History and Memory* 5, 1 (Spring–Summer 1993): 42–86.

8. That the Zeppelin relied on some type of psychological symbolism is indisputable, but whether one can derive a mass psychology from this is open to question. Contemporaries of the airship craze, from Freud to the futurist Marinetti, described the sexual implications of the machine in a number of ways. Some speculated that the Zeppelin worked best as a kind of phallic symbol of unity for a nation suffering from arrested development and inferiority complexes; others preferred the notion of a womb symbol, protecting its children-passengers. Psychohistorical approaches do not adequately explain the phenomenon, for not only do they overlook certain structural factors, but they also make the nuances invisible: Germans may have agreed on viewing Zeppelin as a hero, but their reasons for doing so varied widely. Karl Clausberg, *Zeppelin: Die Geschichte eines unwahrscheinlichen Erfolges* (1979; reprint, Augsburg: Weltbild, 1989), chap. 10. In *The Interpretation of Dreams* (1904), Freud wrote, "As a very recent symbol of the male organ I may mention the airship whose employment is justified by its relation to flying and also occasionally by its form." Douglas Robinson, letter to the author, 3 February 1992.

9. Fritzsche, *Nation*, 18–19; Peter Fritzsche, *Reading Berlin, 1900* (Cambridge: Harvard University Press, 1996), 3–8.

10. Bernd Jürgen Warneken, "Zeppelinkult und Arbeiterbewegung," *Zeitschrift für Volkskunde* 80 (1984): 62–65, 73.

11. BA, R5/3839; cf. Fritzsche, *Nation*, 26, 31. The conservative *Frankfurter Zeitung* (4th morning ed., 21 August 1908) interpreted the reversal of position as a sign that Social Democracy lacked the power to control at least this one example of German work and inventive spirit.

12. Henry J. Schmidt, *How Dramas End: Essays on the German Sturm und Drang, Büchner, Hauptmann, and Fleisser* (Ann Arbor: University of Michigan Press, 1992), 2.

13. Louis L. Snyder, *Roots of German Nationalism* (Bloomington: Indiana University Press, 1978), 45–50.

14. Schmidt, 4–5.

15. Bernd Klagholz, *Der Tag von Echterdingen* (Leinfelden-Echterdingen: Stadtarchiv, 1998), 90–100.

16. Georg Hacker, *Die Männer von Manzell* (Frankfurt a.M.: Societäts-Verlag, 1936), 101–6.

17. Hermann Stegemann, "Errinerungen aus meinem Leben und meiner Zeit" (1930), reprinted in *Deutsche Sozialgeschichte, 1870–1914*, ed. G. A. Ritter and J. Kocka (Munich: Beck, 1983), 101–2; LZA, 04/233, letter of Mainz craftsman Schlitz to Frau Uhland (wife of Zeppelin associate Ernst Uhland), August 1909.

18. LCM, Hildebrandt papers, box 34, offprint of *Preussischen Jahrebüchern* 142, 1 (1910): 113.

19. Adolf Saager, ed., *Zeppelin: Der Mensch—Der Kämpfer—Der Sieger* (Stuttgart: Robert Lutz, 1915), 13, 165–66, 173–74, 205–6, 230–32; cf. Clausberg, chap. 12; Fritzsche, *Nation*, 17–21 and 225 n. 21. Other poems, including some in local dialect, are contained in the following Zeppelin archival files: LZA, 03/007, 03/008, 03/029, 04/75, 04/187, 04/210, and 04/236.

20. *Zeitschrift des VDI*, 1908, 1549–52.

21. BA, R43F/1335, R5/3839; BDM, N94 (Rudolph Diesel papers), box 14; cf. Fritzsche, *Nation*, 33; Clausberg, 57–59.

22. SWB, Pr22/31, *Aufruf!* list, 6 August 1908.

23. BA, R5/3838, unidentified clipping, 10 August 1908; *Berliner Tageblatt*, 11 August 1908; *National-Zeitung*, 15 August 1908.

24. *Vorwärts*, 9 August 1908; cf. BA, R5/3838, *Hannoverscher Courier*, 9 August 1908; *Hamburger Zeitung*, 11 August 1908; Henry Cord Meyer, *Count Zeppelin: A Psychological Portrait* (Auckland, New Zealand: Lighter-Than-Air Institute, 1998), 44–45.

25. *Die Hilfe*, 16 August 1908; *Reichstag Berichte, 166. Sitzung*, 23 November 1908, 5664.

26. Fritzsche, *Reading*, 226–29.

27. Bernhard von Bülow, *Memoirs of Prince von Bülow III: The World War and Germany's Collapse, 1909–1919* (Boston: Little, Brown, 1932), 50–51; Hans G. Knäusel, *LZ 1, der erste Zeppelin: Geschichte einer Idee, 1874–1908* (Bonn: Kirschbaum, 1985), 198.

28. *Wer ist's?* 1906 and 1909; *Meyers Konversation Lexikon*, 1909, 960; cf. 1909–10, 1099.

29. Douglas Robinson, *The Zeppelin in Combat*, 3d ed. (Seattle: University of Washington Press, 1980), 17, 19, 28; Henry Cord Meyer, *Airshipmen, Businessmen, and Politics, 1890–1940* (Washington, D.C.: Smithsonian, 1991), 21–52. The admiration Count Zeppelin enjoyed at home and abroad was not without its problems. Convinced, for example, that he had the right solution as far as airships went, the count at first remained deaf to advice from physicists and engineers such as Ludwig Prandtl that he modify the shape of his machines. This was not the first time the

count would defy authority, scientific or military, and the incident is a telling illustration of his character and of his apparent tendency to thumb his nose at power.

30. Joseph Campbell, *The Hero with a Thousand Faces* (1949), quoted in Daniel J. Boorstin, *The Image* (New York: Atheneum, 1962), 51.

31. LZA, 03/025, provides an overview of the birthday celebrations. A rough parallel to the *Niebelungenlied* is drawn in Arthur Huebner, *Das Lied vom Zeppelin* (Berlin: Pharus-Verlag, 1913).

32. Hans Rosenkranz, *Graf Zeppelin: Die Geschichte eine abenteuerlichen Lebens* (Berlin: Ullstein, 1931), 75.

33. ACD, Stinnes papers, 14/6, comments by President of the Prussian Parliament Count von Schwerin-Löwitz written on the transcription of a speech by Count Zeppelin; cf. Rolf Italiaander, *Ferdinand Graf von Zeppelin* (Konstanz: Stadler, 1986), 146.

34. H. L. Nieburg, *Culture Storm: Politics and the Ritual Order* (New York: St. Martin's Press, 1973), 90.

35. *Deutsche Illustrierte Zeitung* 4 (1909): 12. One admirer went so far as to carve a nutcracker in the count's likeness from the wood of the tree into which LZ 4 had slammed before exploding.

36. *Schlesische Zeitung,* 25 June 1913. During a visit to Leipzig, the count reportedly said, "That the people make me a source of their joy was not my intent. I had not planned it. This is why this ovation impresses me."

37. Clausberg, 74–91, 105–7.

38. NASM, A3G-309040-01; Paul Matssdorf, ed. *Jugend- und Volksbühne 68/69: Graf Zeppelin* (Leipzig: Arwed Strauch, n.d.). Claiming that it had the support of not only Count Zeppelin and his associates but also more than one hundred scientists from Germany, Austria-Hungary, and Switzerland, the Zeppelin Union had set about raising funds for the establishment of a Zeppelin museum and sponsorship of a polar exploration by airship. Eventually, despite the apparently harmless goals of the organization, the count put an end to its use of his name.

39. LZA, 04/192. Other such organizations included the "Kepler-Bund," devoted to the study of natural sciences; Clausberg, 112.

40. LZA, 03/016 and 03/020; cf. Clausberg, 110–13.

41. LZA, 20/019, Magnus Schwantje, "Das Luftschiff als Symbol des Tierschutzes," *Ethische Rundschau* 3, 7–8 (July–August 1914): 109–10.

42. WSU, Wright papers, box 6, file 188, A. Hildebrandt, *Die Brüder Wright* (Berlin: Elsner, 1909), 3; *The Papers of Wilbur and Orville Wright*, vol. 2, ed. Marvin W. McFarland (New York: Arno, 1972), 964.

43. *Deutsche Illustrierte Zeitung* 49 (1909): 5.

44. Fritzsche, *Nation*, 20–21, 226 n. 32.

45. BA, NL 42 (Külz), vol. 155; events surrounding the canceled and rescheduled flight of the *Sachsen* to Zittau in the summer of 1913 prompted letters from

Count Zeppelin and two of his assistants to clarify the situation. LZA, 04/186, letter of Eckener to Count Zeppelin, 13 May 1912; *Bonner Zeitung*, 6 August 1909; cf. *General Anzeiger*, 20 September 1909; cf. Ernest August Schröder, "Von den Wehen in der Geburtstunde des Luftverkehrs," *Beiträge zur Geschichte von Stadt und Stift Essen* 104 (1991–92): 110–12.

46. SWB, Pr31/391, *Bedingungen für Besuchsfahrten mit dem Luftschiff "Charlotte."*

47. Alfred Colsman, *Luftschiff voraus!* (1933; reprint, Munich: Zuerl, 1983), 71–72, 75–76.

48. LZA, 04/202, *Saarbrücker Zeitung*, 21 November 1912; *Deutsche Rundschau*, 9 November 1912.

49. LZA, 04/236, *Krefelder Zeitung*, 11 May 1911.

50. Gerhard Fieseler, *Meine Bahn am Himmel* (Munich: Bertelsmann, 1979), 26–27.

51. Ludwig Laber, *Zeppelin über München* (Munich: Senioren-Buch Verlag, 1984), 8–10; cf. Hacker, 75, 137; *Deutsche Illustrierte Zeitung* 26 (1913): 7. The 1913 Cologne carnival parade featured a "Zeppelin-Pegasus" float that was a great favorite with the crowd.

52. HAStK, Trimbhorn papers, Best 1256, vol. 252, letter of Cologne mayor Walraff to War Minister von Heeringen, 23 December 1910.

53. Günter Schmitt and Werner Schwipps, *Zwanzig Kapitel frühe Luftfahrt* (Berlin: Transpress, 1990), 176–77.

54. Velimir Chlebnikov, "Muster für Wortneuerungen in der Sprache," in *Reisen im Luftmeer*, ed. Karl Riha (Munich: Hanser, 1983), 263–65.

55. LZA, 04/202, *Hannoversche Anzeiger*, 11 July 1912. The *Viktoria-Luise* (LZ 11) was one of eleven German rigid airships to bear a name.

56. C. J. Wells, *German: A Linguistic History to 1945* (Oxford: Clarendon Press, 1985), 345, 374.

57. Rolf Italiaander, *Ein Deutscher namens Eckener* (Konstanz: Stadler, 1981), 91; Svante Stubelius, *Airship, Airplane, Aircraft: Studies in the History of Terms for Aircraft in English* (Göteborg: Lund, 1958), 58–172.

58. Stubelius, 119. Evidence suggests that the German term was adopted into English, in a peculiar reversal of sociolinguistic influence.

59. Wells, 378–79.

60. John Barry, *Technobabble* (Cambridge: MIT Press, 1991), 1.

61. LZA, 04/210, *Die Zukunft*, 15 August 1908, 237–50; cf. Dieter Zastrow, *Entstehung und Ausbildung des französischen Vokabulars der Luftfahrt mit Fahrzeugen: "Leichter als Luft" (Ballon, Luftschiff) von den Anfängen bis 1910* (Tübingen: Niemeyer, 1963), 239–42.

62. BDM, Poeschel papers, *Zeitschrift des allgemeinen deutschen Sprachvereins*, 24, 5 (May 1909): 138; 24, 11 (November 1909): 332; 29, 1 (January 1914): 9–10.

63. Quoted in George Thomas, *Linguistic Purism* (London: Longman, 1991), 107.

64. Contemporary German historiography often limits itself to the confines of Bismarck's Germany, ignoring the proximity of Austria and Germanophone Switzerland; Richard J. Evans, *Rethinking German History: Nineteenth Century Germany and the Origins of the Third Reich* (London: Allen & Unwin, 1987), 8, 15.

65. *Neue Zürcher Zeitung*, 25 September 1907.

66. Lutz Tittel, *Die Fahrten des LZ 4, 1908* (Friedrichshafen: Zeppelin Museum, 1983), 22–27; Hacker, 65–70.

67. BA, R5/3830, telegram, 3 July 1908.

68. *Tägliche Rundschau*, 4 July 1908; cf. *Die Woche* 28 (11 July 1908): 1191–94. Sandt sought to recreate some of his impressions of the flight in a popular novel, *Cavette!* (Minden: Bruns, 1908).

69. Emil Sandt, "Die Schweizerfahrt des Grafen Zeppelin am 1. Juli 1908," in *Buch des Fluges*, vol. 3, ed. Hermann Hoernes (Vienna: Szelinski, 1912), 410.

70. Alfred Liebi, *Das Bild der Schweiz in der deutschen Romantik* (Bern: Haupt, 1946), 77.

71. Jean Paul [J. P. Friedrich Richter], *Des Luftschiffers Giannozzo Seebuch* (1800; reprint, Frankfurt a.M.: Suhrkamp, 1975); W. Behringer and C. Ott-Koptschalijski, *Der Traum vom Fliegen: Zwischen Mythos und Technik* (Frankfurt a.M.: Fischer, 1991), 333–39.

72. Liebi, 129.

73. Hacker, 66. Georg Hacker, who was the count's first pilot, described the aerial view of the Rhinefalls as "entirely different" from what riverbank spectators could see.

74. Warneken, 72; Hugo Maeder, "Leichter als Luft," in *Heimatbuch Dübendorf 1979* (Dübendorf: n.p., 1983), 77–94; cf. Thomas Schärli, "Vor 75 Jahren: Zeppelin über Zollikon," *Zolliker Jahrheft, 1983* (Zollikon: n.p., 1983), 69–75.

75. *Neues Wiener Tagblatt*, 26 September 1907. The airplane was described as the future automobile of the air, while the airship would be called upon to take on grander functions, including military ones. *Wiener Luftschiffer Zeitung*, 8, 1 (January 1909). The attempt to form a Genevan "Swiss League for Air Travel" eventually failed for lack of a clear goal.

76. *Wiener Luftschiffer Zeitung*, 11, 10 (May 1912); Walter Bock, "Die publizistische Auseinandersetzung mit der Luftfahrt, untersucht von den Anfängen der Luftfahrt in Österreich bis zur wirtschaftlichen Notwendigkeit der Zivilluftfahrt heute" (Ph.D. diss., University of Vienna, 1985), 23–39. Until his death in 1910, Vienna's influential mayor Karl Lueger maintained that the city should not support aeronautics in any form.

77. *Wiener Luftschiffer Zeitung*, 5, 2 (February 1906): 39–40; cf. Viktor Silberer, *Der heutige Stand der Luftschiffahrt* (Vienna: Allgemeine Sport Zeitung, 1900,

1905, 1913). In each of his presentations, Silberer attacked or commented skeptically on Count Zeppelin and his work.

78. KA, B/562 I, 5; KA, B/562 III, 1906 manuscript, 12. Claiming that "the time of the captive balloon is over," Hoernes promoted the establishment of aviation courses for the Austro-Hungarian imperial armed forces. Following a visit to Friedrichshafen where he met with the count, he returned convinced of the potential military value of the rigid airship.

79. *Neue Freie Presse,* 29 September 1907; *Illustriertes Wiener Extrablatt,* 16 October 1907; cf. *Ostdeutsche Rundschau* (Vienna), 1 August 1900; *Die Zeit* (Vienna), 7 July and 22 September 1900. Tabloids would promptly publish "instant pictures" *(Momentphotographien)* on the cover to ensure sales; this was the case for the LZ 1 flights.

80. *Illustriertes Wiener Extrablatt,* 5 and 6 August 1908.

81. Ongoing reports in *Wiener Luftschiffer Zeitung,* 1909–13; KA, Mannsbarth papers, B/570/1, unpublished manuscript on lighter-than-air craft.

82. OSA, Fach 83, Karton 1, letter of Ladislav von Szögyény-Marish to Minister Aerenthal, 13 August 1908.

83. Rudolf Martin, *Das Zeitalter der Motorluftschiffahrt* (Leipzig: Theod. Thomas, 1907), 79.

84. *Allgemeine Automobil Zeitung,* 12 June 1910, 20.

85. *Neue Freie Presse,* 23 October 1909.

86. Urban-Raghenfred, *Luftschiffahrt und Flugwesen* (Vienna: Verlag der Akademischen Flugwissenschaften Vereinigung, 1912), 5, 8. Jakob Degen (a nineteenth-century Viennese watchmaker), Johann von Laicharding (an Innsbruck scientist), and David Schwarz were often acknowledged for their efforts. Schwarz's origins were also the subject of argument; some claimed he was Hungarian rather than Austrian, while others interpreted his decision to fly in Russia and then in Berlin as a rejection of Austria-Hungary. In acknowledging the value of the nonrigid-airship types that the German August von Parseval had built, the Austrian literature of the period often reclaimed as an Austrian hero the pioneer Paul Hänlein, who had first used the method of internal ballons that Parseval adopted.

87. Reinhard Keimel, *Österreichs Luftfahrzeuge* (Graz: Weishaupt, 1981), 41–65. The Austro-Hungarian army tried out four different airships with the idea of selecting the best design before placing further orders.

88. *Leipziger Abendzeitung,* 27 August 1908.

89. KA, Mannsbarth papers, B/570, clipping from *Verkehr* 41 (1950).

90. BA, R5/3839, *Leipziger Abendzeitung,* 27 August 1908.

91. *Illustrierte Kronen Zeitung,* 8 July 1913.

92. *Österreische Flugzeitschrift* 7, 12 (30 June 1913): 273–77.

93. *Wiener Zeitung,* 10 June 1913.

94. *Die Wahrheit,* 13 June 1913.

95. Leopold Blasel, "Ein Gruss vom deutschen Volk," *Neues Wiener Tagblatt,* 10 June 1913, 13. The wording may have been borrowed from the 1909 poem "Zeppelin," reprinted in Hugo von Hoffmannsthal, *Prosa II* (Frankfurt a.M.: Fischer, 1959), 355–56.

96. *Neues Wiener Tagblatt,* 10 June 1913.

97. *Arbeiter Zeitung,* 9 and 10 June 1913.

98. *Wiener Abendpost,* 10 June 1913.

99. Hugo Eckener, *Graf Zeppelin: Sein Leben nach eigenen Aufzeichnungen und persönlichen Errinerungen* (Stuttgart: Cotta, 1938), 47–62; cf. Meyer, *Airshipmen,* 26.

100. Viktor Silberer, ed., *Warnende Stimmen in Bezug auf Zeppelin-Ballons* (Vienna: Seidel, 1914).

101. Hermann Hesse, *Luftreisen* (Frankfurt a.M.: Insel, 1994), 10–11; Hans G. Knäusel, *Sackgasse am Himmel* (Bonn: Kirschbaum, 1988), 25; Italiaander, *Ferdinand Graf von Zeppelin,* 72, 139.

102. Jürgen Eichler, *Luftschiffe und Luftschiffahrt* (Berlin: Brandenburgisches Verlagshaus, 1993), 105–6.

103. NASM, A3G-309040-01, Office of the U.S. Naval Attaché in Berlin, translation of Müller-Breslau, "On the History of the Zeppelin Air Ship," 1924.

104. Eichler, 106–8.

105. *Neues Wiener Tageblatt,* 30 July 1911, "Spazierfahrt in der Luft," reprinted in Hesse, 10–11.

106. "Projeckt eines Lenkballons mit Stahlblechhülle des Oberleutnants v. Wallborn," *Flug- und Motor-Technik,* 25 November 1909, 27–29.

107. "Das Strakasche Luftschiff," *Flug- und Motor-Technik,* 25 October 1909, 22–23; Arthur Sauer and Jakob Haw, *Aëroplan System Haw-Sauer* (Trier: Lintz, 1910).

108. Eichler, 131–33; Dorothea Haaland, *Der Luftschiffbau Schütte-Lanz Mannheim-Rheinau (1909–1925)* (Mannheim: Institut für Landeskunde und Regionalforschung, 1987); Douglas Robinson, *Giants in the Sky: A History of the Rigid Airship* (Seattle: University of Washington Press, 1973), 82. Robinson suggests that Schütte-Lanz's limited success may have been due not only to the public enthusiasm for Zeppelin airships but also to the fact that Johann Schütte made himself unpopular with government authorities.

109. *Zeitschrift für Flugtechnik und Motorluftschiffahrt* 20 (1910): 209, 262.

110. *Die Woche,* 3 October 1908, 1745–48.

111. Colsman, 47, 99–101; cf. Meyer, *Count Zeppelin,* 87.

112. Stefan Zyma, *Karl Maybach and His Work* (Düsseldorf: VDI, 1995), 3. Colsman worked with Wilhelm Maybach and his son Karl to establish the Luftfahrzeug-Motorenbau Gesellschaft in 1909. Although Wilhelm, who had left Daimler, was legally bound not to assist any competitive endeavor for three years, no such contract bound his son, who patented his own engine improvements.

113. Heike Vogel, *"Suche ein nettes Zimmer"*: *Die Zeppelin-Wohlfahrt GmbH und der Wohnungsbau in Friedrichshafen* (Friedrichshafen: Zeppelin Museum, 1997).

114. August Maier, "Die Anfänge des Christlichen Metallarbeiterverbands," ed. Georg Wieland, *Schriften des Vereins für Geschichte des Bodensees und seiner Umgebung* 114 (1996): 91–95.

115. On the unhappy experiences of Zeppelin with Taylorism and other mass-production principles, see Jeannine Zeising, "Die 'Tempomacher': Die halbherzige Karriere des Taylorismus im Luftschiffbau," in *Zirkel, Zangen, und Cellon: Arbeit am Luftschiff*, ed. Wolfgang Meighörner (Friedrichshafen: Zeppelin Museum, 1999), 97–108.

116. Claude Dornier, *Aus meiner Ingenieurlaufbahn* (Zug, Switzerland: n.p., 1966), 57.

117. Jürgen Beibler, "Graf Zeppelin und Ludwig Rüb: Die Anfänge des Flugzeugbaus in Friedrichshafen," in *Der Graf, 1838–1917*, ed. Wolfgang Meighörner (Friedrichshafen: Zeppelin Museum, 2000), 177–91. Although thoroughly convinced that the airship solution was the best, Count Zeppelin had looked into the potential advantages of the airplane as early as 1899, dealing on occasion with early German airplane designers such as Ludwig Rüb, Hans Coler, and Theodor Kober.

118. Dornier, 15; Wolfgang Meighörner, *Alfred Graf von Soden-Fraunhofen* (Friedrichshafen: Zeppelin Museum, 1994), 23–33.

119. Gerhard Hecker, *Walter Rathenau und sein Verhältnis zu Militär und Krieg* (Boppard a.R: Boldt, 1983), 110–11.

120. Eichler, 127.

121. Rosenkranz, 173.

122. *Deutsche Illustrierte Zeitung*, 15 September 1912; 1 June 1913; 28 December 1913.

123. *Deutsche Luftschiffer Zeitung* 22 (1913), 523, reprinted in Eichler, 169. The most notorious of these subsequent accidents was the explosion of the German navy's L 2 during an exercise, with the loss of all hands. The official condolence messages were all framed in terms of giving one's life so that Germany might keep on flying.

124. LCM, Hildebrandt papers, box 34, undated article [ca. 1912]; HHW, 405I-P/4, vol. 2496, letter of the Mittelreheinischer Verein für Luftschiffahrt (Mid-Rhine Airship Association), Mainz, 3 June 1908.

125. BA, Film Archive no. 448, *Eine Luftfahrt Gotha-Düsseldorf* (1912).

126. Imperial law decree of 12 May 1894 and 1906: imperial law for the protection of registered trademarks, 126, quoted in Clausberg, 113.

127. Jennifer Jenkins, "The Kitsch Collections and *The Spirit of Furniture*: Cultural Reform and National Culture in Germany," *Social History* 21, 2 (May 1996): 123–41.

128. Heike Schröder, "Pazaurek und die 'Sammlung der Geschmacksverirrungen,'" in *Die Kunst des Fliegens,* ed. Bodo-Michael Baumunk (Friedrichshafen: Zeppelin Museum, 1996), 49–51; Jean Munon, *L'avion jouet témoin de son temps* (Paris: Aéroport de Paris exhibit catalog, 1979), 3.

129. Jenkins, 137.

130. "Die Frau als Luftschifferin," *Die Woche* 37 (12 September 1908): 1609–11; cf. "Luftschiffer Kleidung," *Deutsche Zeitschrift für Luftschiffahrt* 16, 7 (6 April 1910): 29.

131. BA, NL103/29, minutes of the 2 July 1914 meeting of the airship section of the Deutsche Luftfahrer Verband (German Aviators Association), "Antrag des Fräulein Riotte. . . ."

132. Letter of Elly Heuss-Knapp to Georg Friedrich Knapp, 15 September 1909, in *Elly Heuss-Knapp,* ed. Margaret Vater (Tübingen: Wunderlich, 1961); cf. Heinz Dörr, ed. *Wilhelm Ernest Dörr, 1882–1954* (Überlingen: Stadtverwaltung, 1993), 109–12; Hesse, 19.

133. Wolfgang Schivelbusch, *The Railway Journey* (Berkeley: University of California Press, 1986), 134–49.

134. *Luftfahrt Beiblatt zu den Hamburger Beiträgen,* 2 April 1914.

135. Hans Dominik, "Die Technik des zwanzigsten Jahrhundert," *Die Gartenlaube* 3 (1910): 73–74.

136. *Neue Freie Presse,* 25 December 1909.

137. Reyner Banham, *Theory and Design in the First Machine Age* (1960; reprint, Cambridge: MIT Press, 1980), 11.

138. Tilmann Buddensieg, *Industriekultur: Peter Behrens and the AEG* (Cambridge: MIT Press, 1984), 16.

139. Ibid., 65–66.

140. Quoted in Modris Eksteins, *The Rites of Spring: The Great War and the Birth of the Modern Age* (1989; reprint, London: Black Swan, 1990), 43.

141. Buddensieg, 79.

142. "Mögliche und unmögliche Luftschiffprojeckte," *Motor,* April 1913, 64.

143. Reinicke, 313.

144. Ibid., 314–15.

145. Paul Neumann, *Luftschiffe* (Bielefeld: Velhagen & Klasing, n.d.), 1–2.

146. Raymond Grew, "The Construction of National Identity," in *Concepts of National Identity: An Interdisciplinary Dialogue,* ed. Peter Boerner (Baden-Baden: Nomos, 1986), 31–44.

147. Eksteins, 322.

148. Gustav Radbruch, *Biographische Schriften,* ed. Günter Spendel (Heidelberg: C. F. Müller, 1988), 168.

149. Michael Nerlich, *Ideology of Adventure* (Minneapolis: University of Minnesota Press, 1987).

150. Leo Braudy, *The Frenzy of Renown: Fame and Its History* (New York: Oxford University Press, 1986), 19–22.

151. David Blackbourne and Geoff Eley, *The Peculiarities of German History* (New York: Oxford University Press, 1984), 144–47; cf. Evans, 23–54.

Chapter Three. Zeppelin Myth and Reality in the Great War

1. Jeffrey T. Schnapp, "Propeller Talk," *Modernism/Modernity* 1, 3 (1994): 153–78.

2. Helmut Reinicke, *Aufstieg und Revolution: Über die Beförderung irdischer Freiheitsneigungen durch Ballonfahrt und Luftschwimmkunst* (Berlin: Transit, 1988), 128.

3. Michael Paris, *Winged Warfare: The Literature and Theory of Aerial Warfare in Britain, 1859–1917* (Manchester: Manchester University Press, 1992), chap. 2; cf. Claus Ritter, *Kampf um Utopolis oder die Mobilmachung der Zukunft* (East Berlin: Verlag der Nation, 1987).

4. Hans Waldemar von Herwarth, *Unser Luftreich-unsere Zukunft* (Berlin: Continent, 1912), 11.

5. Alfred Gollin, *No Longer an Island: Britain and the Wright Brothers, 1902–1909* (Stanford, Calif.: Stanford University Press, 1984), 336–38.

6. David Lloyd George, *War Memoirs, 1914–1915* (Boston: Little, Brown, 1933), 30–31, 33.

7. Alfred Gollin, *The Impact of Air Power on the British People and Their Government, 1909–14* (Stanford, Calif.: Stanford University Press, 1989), 13–14.

8. Ibid., 53–55.

9. Ibid., 58–59.

10. Ibid., 223–27.

11. Ibid., 254.

12. "The Next War in the Air," *Pearson's Magazine*, July 1913, 43–44; BA, R5/3836, *Hamburger Fremdenblatt*, 6 June 1914; *Kieler Zeitung Morgenblatt*, 1 July 1914; cf. *Die Umschau*, 8 February 1913, 131; 1 January 1916, 7.

13. Quoted in Gollin, *Impact*, 227.

14. *Deutscher Illustrierte Zeitung* 7 (1913): 3.

15. *Berliner Lokal Anzeiger*, 4, 11, and 13 July 1908; cf. Paris, 125, 127; Robin Higham, *The British Rigid Airship, 1908–1931: A Study in Weapons Policy* (London: G. T. Foulis, 1961), 10–11, 56–57.

16. "Die englische Angst: Ein Beitrag zur grossbritannischen Luftschiffpanik," *Motor*, April 1913, 44–49.

17. BA/MA, PH9–V/51, reports concerning the Lunéville incident, April 1913; Ministère des affaires étrangères, *Documents diplomatiques français (1871–1914)*, vol. 6 (Paris: Imprimerie nationale, 1933), 231, 235; cf. Johannes Lepsius et al., *Die*

Grosse Politik der Europäischen Kabinette, 1871–1914, vol. 39 (Berlin: DVPG, 1926), 281–303.

18. LZA, 04/290, report of Zeppelin mechanic Ludwig Marx, 7 April 1913; *La conquête de l'air* 8 (April 1913): 122; Raymond Poincaré, *The Memoirs of Raymond Poincaré,* vol. 2, trans. Sir George Arthur (New York: Doubleday Doran, 1928), 50–51.

19. *Deutsche Luftschiffer Zeitung* 18, 3 (4 February 1914): 65.

20. Service historique de l'Armée de terre (Paris), 7N 1112-1, report no. 75 of the French military attaché in Berlin, 23 April 1913; Eugen Weber, *The Nationalist Revival in France, 1905–1914* (Berkeley: University of California Press, 1968), 120–23; Agnès Bouhet, "L'affaire Saverne," *Guerres mondiales et conflits contemporains* 173 (1994): 5–17. Saverne (Zabern) was an Alsatian town that housed a Prussian regiment; a young officer there ordered a charge against unarmed demonstrators in November 1913, precipitating a governmental crisis in Germany.

21. Stephen Kern, *The Culture of Time and Space, 1880–1918* (Cambridge: Harvard University Press, 1983), 249–51.

22. BA, R5/3829, minutes of meeting between Count Zeppelin and imperial officials, 19 December 1906; *Der Tag,* 25 June 1909.

23. *Die Umschau* 39 (22 September 1906): 761–66.

24. Heinrich Thalmann, "Erster Zeppelin-Überflug der Pfalz und nationale Begeisterung im Jahre 1908," *Pfälzischer Heimat* 43 (1992): 179.

25. LZA, 04/165, letter of Ernest Dalle (Bremen) to Count Zeppelin, 7 August 1908. Dalle held the view that humans were not meant to fly and warned the count that the destruction of LZ 4 in Echterdingen was a sign of God's opposition to his war designs.

26. A. Stelling, *12000 Kilometer im Parseval,* quoted in Jürgen Seifert, *Die Luftschiffwerft und die Abteilung Seeflugzeugbau der Luft-Fahrzeug-Gesellschaft in Bitterfeld (1908–1920)* (Bitterfeld: Rat des Kreises Bitterfeld, 1988), 24.

27. "Die Flugtechnik im Dienste des Krieges," *Flug- und Motor-Technik,* 25 October 1909, 1–8.

28. "Die Kölner Übungsfahrten," *Flug- und Motor-Technik,* 25 November 1909, 18–20.

29. Martin, 80–86; cf. Paris, 128–29; Gollin, *Impact,* 5–7, 13–14, 58.

30. Ritter, 263.

31. BA/MA, PH9-V/41, airship battalion position paper, June 1910; Dorothea Haaland, *Der Luftschiffbau Schütte-Lanz Mannheim-Rheinau (1909–1925): Die Geschichte einer innovativen Idee als zeitlich räumlicher Prozess* (Mannheim: Institut für Landeskunde und Regionalforschung, 1987), 35–37.

32. Douglas Robinson, *The Zeppelin in Combat,* 3d ed. (Seattle: University of Washington Press, 1980), 18–28.

33. Jürgen Eichler, *Luftschiffe und Luftschiffahrt* (Berlin: Brandenburgisches Verlagshaus, 1993), 127.

34. *Reichstag Berichte, 177. Sitzung,* 5 December 1908, 6011; *178. Sitzung,* 7 December 1908, 6035; *179. Sitzung,* 9 December 1908, 6095.

35. David Herrmann, *The Arming of Europe and the Making of the First World War* (Princeton: Princeton University Press, 1996), 147–98.

36. Helmut Reinicke, "Begeisternde Arbeit und schwereloser Aufstieg vor 1914," in *Vom Wert der Arbeit: Zur literarischen Konstitution des Wertkomplexes "Arbeit" in der deutschen Literatur (1770–1830),* ed. Harro Segeberg (Tübingen: Niemeyer, 1991), 297.

37. Pascal Vennesson, *Les chevaliers de l'air: Aviation et conflits au XXe siècle* (Paris: Presses de sciences Po, 1997), 50–51; Robert Wohl, *A Passion for Wings: Aviation and the Western Imagination, 1908–1918* (New Haven: Yale University Press, 1994), 97.

38. G. von Arnauld de la Perière, *Prinz Heinrich von Preussen: Admiral und Flieger* (Herford: Koehler, 1983), 55–56; John H. Morrow Jr., *Building German Air Power, 1909–1914* (Knoxville: University of Tennessee Press, 1976), 48–49.

39. Morrow, 57–59.

40. LHK, Oberpräsidium (403), no. 12122; cf. HHW, 405/I-P/4, vol. 2601.

41. "Die berliner Flugwoche," *Flug- und Motor-Technik,* 10 November 1909, 1–14.

42. LHK, Oberpräsidium (403), no. 12123, letter of RMI to Oberpräsidium, 5 February 1913. The government declined to help develop the local *Veeh* dirigible project on the grounds that funding from the 1912 national drive was reserved exclusively for airplanes.

43. LCM, Hildebrandt papers, box 18, letter of Colsman to Hildebrandt, 25 May 1912.

44. Letter of Lerchenfeld to Hertling, 15 May 1912, in Georg Graf von Hertling and Hugo Graf von und zu Lerchenfeld, *Briefwechsel Hertling-Lerchenfeld, 1912–1917,* vol. 1, ed. Ernst Deuerlein (Boppard: Boldt, 1973), 161.

45. Gerd Fritz Leberecht, *Luftfahrten im Frieden und Kriege* (Berlin: Leonard Simion, 1913), 248; BA, R5/3829, minutes of meeting between Count Zeppelin and imperial officials, 19 December 1906; *Reichstag Berichte, 21. Sitzung,* 18 March 1907, 568; R5/3832, letter of Count Zeppelin to RMI, 18 July 1909.

46. *Die Militärluftfahrt bis zum Beginn des Weltkrieges 1914,* vol. 1, 2d ed. (Frankfurt a.M.: Mittler, 1965), 86, quoted in Eichler, *Luftschiffe,* 130.

47. SHAA, A161, dossier 2/4, undated report [1913–14], "Rôle offensif des dirigeables."

48. Quoted in Reinicke, "Begeisternde Arbeit," 300.

49. BA, R5/3832, letter of Count Zeppelin to RMI, 2 June 1909; R5/3833, RMI memorandum, 29 April 1910; letter of Zeppelin company to AA, 8 May 1912; RMI memorandum, 19 August 1912; minutes of RMI meeting with Count Zeppelin, 29 July 1912; R5/3834, letter of War Ministry to Chancellor's Office, 1 November 1913;

cf. Henry Cord Meyer, *Count Zeppelin: A Psychological Portrait* (Auckland, New Zealand: Lighter-than-Air Institute, 1998), 74–75.

50. "Die deutsche Luftmacht," *Motor,* July 1913, 61–62.

51. Wolfgang Meighörner, *Alfred Graf von Soden-Fraunhofen* (Friedrichshafen: Zeppelin Museum, 1994), 33.

52. Heinrich Walle, "Das Zeppelinsche Luftschiff als Schrittmacher technologischer Entwicklungen in Krieg und Frieden," in *Militär und Technik: Wechselbeziehungen zu Staat, Gesellschaft, und Industrie im 19. und 20. Jahrhundert,* ed. Roland Foerster and Heinrich Walle (Bonn: Mittler, 1992), 187.

53. BA/MA, Seekt papers, N247/62, "Ansprache des Grafen Zeppelin über die Bedeutung der Luftschiffe im Weltkrieg und ihre Zukunftaugaben"; cf. Meyer, *Count Zeppelin,* 92–97.

54. Rolf Italiaander, *Ein Deutscher namens Eckener* (Konstanz: Stadler, 1981), 123; cf. Hans Knäusel, "Zeppelin und Zeppelinismus," *Trans* 1 (1989): 86.

55. Peter W. Brooks, *Zeppelin: Rigid Airships, 1893–1940* (Washington, D.C.: Smithsonian, 1992), 78–79.

56. Robinson, *Combat,* 25–26; Brooks, 68–69.

57. BA, NL96/3 (Dörr), "Luftschiff starren Systems Gerippe aus Holz." Schütte-Lanz relied on the patents of the civil engineer Carl Huber. W. Rettig, a shipbuilder from Berlin, had proposed a helicoidal wooden airship hull before Huber, but he found no backers.

58. Dorothea Haaland, "Der Luftschiffbau Schütte-Lanz," in *Leichter als Luft-Ballone und Luftschiffe,* ed. Dorothea Haaland et al. (Bonn: Bernard & Graefe, 1997), 207.

59. Peter Kleinheins, ed., *Die Grossen Zeppeline* (Düsseldorf: VDI, 1985), 74–75.

60. Werner Sombart, *Kapitalismus und Krieg* (Munich: Duncker & Humbolt, 1913), quoted in Alex Roland, "Technology and War: The Historiographical Revolution of the 1980s," *Technology and Culture* 34, 1 (1993): 131.

61. Douglas Robinson, *Giants in the Sky: A History of the Rigid Airship* (Seattle: University of Washington Press, 1973), 91.

62. E. A. Pfeiffer, *Fahren und Fliegen: Ein Buch für alle von Auto, Flugzeug, Zeppelin* (Stuttgart: Franckh'sche Verlagshandlung, 1935), 82–83.

63. Peter Kleinheins, *LZ 120 "Bodensee" und LZ 121 "Nordstern": Luftschiffe im Schatten des Versailler Vertrages* (Friedrichshafen: Zeppelin Museum, 1994), 25.

64. Waldemar Koelle, "Marineluftschiffe im Kriege, in Sturm und Not," *Deutsche Rundschau* 205 (December 1925): 245.

65. Robinson, *Combat,* 227–30.

66. SHAA, A163, dossier 4/4, *Extrait de l'étude sur le Zeppelin L.-49* (Paris: Service technique et industriel de l'aéronautique maritime, 1917).

67. Robinson, *Combat,* 317.

68. Wolfgang Meighörner, *Wegbereiter des Weltluftverkehrs wider Willen: Die Geschichte des Zeppelin-Luftschifftyps "w"* (Friedrichshafen: Zeppelin Museum, 1992); cf. Jürgen Eichler, "Die Afrikafahrt des Marineluftschiffes L 59," *Militärgeschichte* 26, 3 (1987): 248–53.

69. Michel Corday, *The Paris Front* (New York: E. P. Dutton, 1934), 9; Gollin, *Impact*, 315.

70. Michele J. Shover, "Roles and Images of Women in World War I Propaganda," *Politics and Society* 5 (1975): 471–73.

71. Eberhard Demm and Tilman Koops, *Karikaturen aus dem ersten Weltkrieg* (Koblenz: Bundesarchiv, 1990), 13.

72. Paul Vincent, *Cartes postales d'un soldat de 14–18* (Paris: Gisserot, 1988), 7; cf. George Mosse, *Fallen Soldiers: Reshaping the Memory of the World Wars* (New York: Oxford University Press, 1990), 128–29.

73. Eberhard Demm, "Propaganda and Caricature in the First World War," *Journal of Contemporary History* 28 (1993): 163–92.

74. Shawn Aubitz and Gail F. Stern, "Ethnic Images in World War I Posters," *Journal of American Culture* 9 (1986): 83–98; Mosse, 136–37.

75. Hubertus F. Jahn, *Patriotic Culture in Russia during World War I* (Ithaca, N.Y.: Cornell University Press, 1995), 49–50.

76. Stéphane Audoin-Rouzeau, *La guerre des enfants, 1914–1918* (Paris: Armand Colin, 1993), 24–65; Karl Hess, *Der I. Weltkrieg und die Schule* (Friedrichshafen: Schulmuseum, 1989).

77. Barbara Jones and Bill Howell, *Popular Arts of the First World War* (New York: McGraw-Hill, 1972), 90, 96–98.

78. Mosse, 137.

79. Marie-Monique Huss, "Pro-Natalism and the Popular Ideology of the Child in Wartime France: The Evidence of the Postcard," in *The Upheaval of War: Family, Work, and Welfare in Europe, 1914–1918,* ed. R. Wall and J. Winter (New York: Cambridge University Press, 1988), 329–67; Audoin-Rouzeau, 124–29;

80. Audoin-Rouzeau, 80–83.

81. Pellerin, *Imagerie d'Epinal,* 87, reprinted in Audoin-Rouzeau, fig. 9, between 96 and 97.

82. SHAA, Oral History Section, interview 311 (van der Dorpe), 19 October 1982.

83. BA/MA, N302/2, Breithaupt papers, manuscript "L 15 wird abgeschossen," 9.

84. Raymond L. Rimell, *Zeppelin! A Battle for Air Supremacy in World War I* (London: Conway, 1984), 117–26.

85. Service historique de l'Armée de terre (Paris), 5N 152, telegram of General Sarrail to French War Ministry, 5 May 1916; Douglas Robinson, "Zeppelin Intelligence," *Aerospace Historian* 21, 1 (March 1974): 3.

86. SHAA, A163, dossier 4/4, letters to Flight Engineer Paul Schlick; cf. Rimell, 129; Robinson, "Zeppelin Intelligence," 4–5.

87. Robinson, *Combat*, 286–98.

88. SHAA, Oral History Section, interview 58 (Renaitour), 13 January 1977; interview 180 (Cournot), 20 December 1979.

89. Robert Colin, quoted in Bruno Théveny, "Treize Zeppelins sur Londres," *CPC* 124 (November–December 1988): 11.

90. James Munson, ed., *Echoes of the Great War: The Diary of Reverend Andrew Clark, 1914–1919* (New York: Oxford University Press, 1985), 47, 86.

91. Rimell, 54.

92. Quoted in *The Imperial War Museum Book of the First World War*, ed. Malcolm Brown (Norman: University of Oklahoma Press, 1993), 220.

93. Munson, entry of 3 October 1916, 161–62; cf. testimonies of Mrs. Holcombe Igleby and Mr. Archie Steavenson, quoted in Brown, 221.

94. Munson, 159–60.

95. Paul Fussell, *The Great War and Modern Memory* (New York: Oxford University Press, 1975), 86.

96. SHAA, Oral History Section, interview 58 (Renaitour), 13 January 1977.

97. Corday, 74–75.

98. Ibid., 47.

99. Ibid., 48.

100. Ottokar Czernin, *In the World War* (London: Cassell, 1919), 101–3.

101. Holcombe Ingleby, *The Zeppelin Raid in West Norfolk* (London: Edward Arnold, 1915), 29. Whereas Czernin thought of the airship as a cigar, a witness who lived through a raid over England recalled it as "the biggest sausage" ever, while another thought of a "church steeple endways."

102. Czernin, 103–4.

103. Jules Poirier, *Les bombardements de Paris (1914–1918): Avions—Gothas—Zeppelins—Berthas* (Paris: Payot, 1930), 198–201; cf. Peter Fritzsche, *A Nation of Fliers: German Aviation and the Popular Imagination* (Cambridge: Harvard University Press, 1992), 40, 229n. The accounts were often taken from police reports or local papers, and assumed that the reader was familiar with the people, or at least with the places mentioned. For example: "One bomb fell on a one-story building. The Petitjean family and a few friends were meeting there on the occasion of Mr. Auguste Petitjean's visit from the front, where he'd been for the past eighteen months. . . . The twelfth bomb fell on a one-story building on Haxo Street, in which the Tédé couple and Mrs. Morillon lived; the former were wounded, she was killed."

104. André Kling, *Bombes et engins explosifs de l'aéronautique allemande* (Paris: Service géographique de l'Armée, 1915), 25–27.

105. Robinson, *Combat*, 179

106. Modris Eksteins, *The Rites of Spring: The Great War and the Birth of the Modern Age* (1989; reprint, London: Black Swan, 1990), chap. 7.

107. Arthur Conan Doyle, "Reprisals," *Times*, 15 October 1915; *Sunday Times*,

6 February 1916; Wohl, 284; Vincent J. Cheng, "The Zeppelin Nights of Ford Madox Ford," *Journal of Modern Literature* 15, 4 (Spring 1989): 595–97.

108. Bernard Shaw, *Collected Letters, 1911–1925*, ed. Dan H. Laurence (New York: Viking, 1985), letter to Siegfried Trebitsch, 9 June 1915, 298. Trebitsch had just published a note in the *Vossische Zeitung* claiming that Shaw was being mistreated.

109. Stanley Weintraub, *Journey to Heartbreak* (London: Macmillan, 1989), 179.

110. William C. Carter, *The Proustian Quest* (New York: New York University Press, 1992), 1–22, 133–204.

111. Marcel Proust, *Remembrance of Things Past*, trans. C. K. Scott Moncrieff (1927; reprint, New York: Random House, 1970), vol. 2, *The Past Recaptured*, 916; cf. Wohl, 284.

112. Dominique David, "L'Air du Temps: A la recherche de l'aviation de Proust," *Revue historique des armées* 4, 2 (1977): 192–93; cf. Wohl, 284–85.

113. Jules Romain, *Men of Goodwill* (New York: Knopf, 1939), vols. 15–16, *Verdun*, 411–14, 418.

114. Carl Heeg, "Ein Zeppelin über der Burg Olbrück im Jahre 1915," *Jahrbuch des Kreises Daun*, 1977, 134–35.

115. Fritzsche, chap. 2.

116. Robinson, *Combat*, 48–49, 65.

117. Ernst Lehmann, *Auf Luftpatrouille und Weltfahrt* (Berlin: Wegweiser, 1936), 57.

118. Hans Floerke and Georg Gärtner, *Deutschland in der Luft voran!* (Munich: Georg Müller, 1915), 5–14.

119. *Kölnische Zeitung* editorial, 21 January 1915, quoted in Robinson, *Combat*, 64; BA, R43F/1336, *Die Luftflotte* 8, 10 (1 October 1915): 89–90, 95–97; Fritzsche, 44.

120. "Ein Luftschiff-Strafexpedition gegen Paris," *Motor*, April–May 1915, 41–42.

121. A. A. Noskoff, "Zur Geschichte des Luftkrieges: Der Zeppelinangriff auf das Russische Hauptquartier in Sieldce am 3. August 1915," *Gasschutz und Luftschutz* 5, 8 (August 1935): 197–99.

122. Audoin-Rouzeau, 67–69.

123. Reinhold Braun, *Deutsche Kriegsbücher für die Schuljugend, 4. Band: Der Krieg in der Luft* (Langensalsa: Julius Belss, n.d.), 7.

124. ACD, Stinnes papers, 14/6; transcript of a speech given by Count Zeppelin in the Prussian Parliament House, Berlin, July 1916; Karl von Einem, *Ein Armeeführer erlebt den Weltkrieg* (Leipzig: Hase & Koehler, 1938), ed. Junius Alter, 15 December 1915 entry, 186–87.

125. Quoted in Ernst Johann, ed. *Innenansicht eines Krieges* (Frankfurt a.M.: Heinrich Scheffler, 1968), 175–76; cf. BA, NL91/2 (Schiffer), memoirs manuscript,

111–12. The conversation reportedly took place on a train on 25 March 1916. National Liberal party member Eugen Schiffer witnessed a similar venting of the count's anger toward England while attending a reception.

126. *Bonner Rundschau*, 20 August 1955; Count Zeppelin's 1908 speech "An das deutsche Volk" was reprinted alongside the writings of Rilke and Fontane in the *Kriegsalmanach 1915* (Leipzig: Insel, 1915), 161–63.

127. BA, NL34/9 (Gok); cf. BA, NL91/4 (Schiffer), memoirs manuscript, 648–50.

128. Theodor Wolff, *Tagebücher, 1914–1919*, vol. 1, ed. Berd Sösemann (Boppard a.R.: Boldt, 1984), 486–87. Wolff added that the count's calls for unrestricted submarine warfare and annexation were proof of his inexperience in politics; letter of Lerchenfeld to Hertling, 21 October 1916, in *Briefwechsel Hertling-Lerchenfeld, 1912–1917*, vol. 2, 762–63.

129. Eulogy reprinted in *Zeppelin Briefe* 20 (February 1992): 1–2.

130. *Der Drahtverhau* 21 (March 1917), cover reprinted in Johann; *Simplicissimus*, 27 March 1917, cover reprinted in Rolf Italiaander, *Ferdinand Graf von Zeppelin* (Konstanz: Stadler, 1986), 182.

131. BA, R43F/1336, "Errinerungen an den grossen Deutschen, den Grafen Zeppelin," *Kauffmännische Blätter* 5–6 (May–June 1917): 33–36; cf. Italiaander, *Zeppelin*, 177–86.

132. Quoted in Johann, 281.

133. Alfred Colsman, *Luftschiff voraus!* (1933; reprint, Munich: Zuerl, 1983), 243.

134. Hermann Hilger, ed., *Unsere Luftkreuzer* (Berlin: Hillger, 1914).

135. *Die Umschau*, 14 August 1915, 652–53.

136. *Zeppeline über England* (Berlin: Ullstein, 1916); Adolf-Victor von Koerber, *Luftkreuzer im Kampf* (Leipzig: Amelang, 1916).

137. BA, R5/3834, *Ansprachen über Erröfnungsfeier der Luftschiffwerft in Zeesen*, 1 April 1916. Schütte's friends encouraged him to correct such mistakes by stressing the contributions he had made to Zeppelin engineering through the transfer of his patents, but on the occasion of a public address, Schütte stated that he bore no grudge and that matters would be rectified once the war was over.

138. W. Herbert, "Zeppeline über den Kampffeldern in West und Ost und Nord," in *Der segelnde Tod*, ed. Josef Karl Brechenmacher (Donauwörth: Ludwig Auer, 1915), 39; cf. press reports in Arnold Jünke, *Zeppelin im Weltkriege* (Leipzig: Abel & Müller, n.d.).

139. Dominick Pisano et al., *Legend, Memory, and the Great War in the Air* (Seattle: University of Washington Press, 1992), 47.

140. LZA, 04/247, *Allgmeine Zeitung* (Chemnitz), 28 January 1915. Airplane pilots were often described as a new breed of knight.

141. Georg Paul Neumann, *In der Luft unbesiegt* (Munich: Lehmanns Verlag, 1923), 272.

142. Paul Schmalenbach, *Die deutschen Marine-Luftschiffe* (Herford: Koehler, 1977), 81.

143. Ibid., 85–86.

144. Brooks, 79; Douglas Botting, *The Giant Airships* (Alexandria, Va.: Time-Life Books, 1981), 60–61.

145. Fritzsche, 55–57; Guy Chapman, ed., *Vain Glory*, 2d ed. (London: Cassell, 1968), 474–76.

146. Koelle, 246–50.

147. Letter of Eckener to Colsman, 3 February 1916, printed in Italiaander, *Ein Deutscher*, 137–38.

148. Ibid.; Schmalenbach, 83.

149. Letters of Eckener to his wife, 6 and 9 August 1918, printed in Italiaander, *Ein Deutscher*, 159–60.

150. Theodor Wolff, ed., *Die Wilhelminische Epoche: Fürst Bülow am Fenster und andere Begegnungen* (Frankfurt a.M.: Atheneum, 1989), 84.

151. Letter of Lerchenfeld to Hertling, 28 January 1916, in Georg Graf von Hertling and Hugo Graf von und zu Lerchenfeld, *Briefwechsel Hertling-Lerchenfeld, 1912–1917*, vol. 2, ed. Ernst Deuerlein (Boppard: Boldt, 1973), 570.

152. ACD, Hugo Stinnes papers, I-220 no. 223/6, public speech of Captain Joly, 29 March 1917.

153. Paul von Hindenburg, *Aus meinem Leben* (Leipzig: Hirzel, 1934), 142; Hans G. Knäusel, *LZ 1, der erste Zeppelin: Geschichte einer Idee, 1874–1908* (Bonn: Kirschbaum, 1985), 74–75; cf. Heinrich Walle, "Das Zeppelinsche Luftschiff als Schrittmacher technologischer Entwicklungen in Krieg und Frieden," in *Militär und Technik: Wechselbeziehungen zu Staat, Gesellschaft, und Industrie im 19. und 20. Jahrhundert*, ed. Roland G. Foerster and Heinrich Walle (Herford: Mittler, 1992), 185. An 1897 diary entry suggests that Count Zeppelin recognized the possibility that the airplane would eventually fly and surpass the airship.

154. Robinson, *Combat*, 65; "The Achievements of the Zeppelins," published anonymously in the *Stockholm Dagblad*, 19 March 1916 (reprint, London: Unwin, n.d.).

155. Fritzsche, 55–58; cf. Robinson, *Giants*, chaps. 5 and 6. Indeed, the airshipmen were under strict orders to avoid pushing the flight envelope of the machine, a fact that the Allies ignored after the war when they tested airships based on the German machines, thus incurring accidents.

156. Letters of Eckener to his wife, 25 June, 8, 10, and 12 December 1917, 12 February 1918, printed in Italiaander, *Ein Deutscher*, 147, 155–57.

157. Hans von Schiller, "In memoriam," *Luftfahrt (Deutsche Zeitschrift für Luftschiffahrt)*, 7 August 1928, 225–26.

158. SWA, F25/40, *Kölnische Zeitung*, undated clipping [March 1919]. The original brochure was entitled "Tirpitz der Totengräber der deutschen Flotte" and

was authored by a retired captain "Persius." The publication built its case by quoting from foreign news reports of the summer of 1918 on the efficiency of the Maybach airship engines. The "counter brochure," entitled "Zeppelin, Maybach Motoren, und Persius," was prepared by Zeppelin manager Alfred Colsman and engine manufacturer Karl Maybach.

159. Colsman, 207–27; cf. Knäusel, "Zeppelinismus," 92–93.

160. Henry Cord Meyer, *Airshipmen, Businessmen, and Politics, 1890–1940* (Washington, D.C.: Smithsonian, 1991), 135–38.

161. BA, R5/3854; *Die Rundschau* 15 (15 August 1916): 141.

162. Josef Jurinek, "Wirtschaftlichkeit des Luftverkehrs der Zukunft," *Weltwirtschaft* 8, 8–9 (1918).

163. BA, NL96 (Dörr), U.S. negotiations for planned delivery of the *Lion* (airship L 72 code name). L 72 eventually ended up in France as the *Dixmude*.

164. SML, Barnes Wallis papers, B2/3, folder 2, Captain Heinen's surrender of L 71, July 1920.

165. Lehmann, 214–15.

166. H. Bruce Franklin, *War Stars: The Superweapon and the American Imagination* (New York: Oxford University Press, 1988), 72–75.

167. Robin Higham, "The Peripheral Weapon in Wartime: A Case Study," in *Naval Warfare in the Twentieth Century, 1900–1945* (London: Croom Helm, 1977), 90.

168. Robinson, *Combat*, chaps. 12, 16, 21.

169. Trudi Tate, "The Culture of the Tank, 1916–1918," *Modernism/Modernity* 4, 1 (January 1997): 69–87.

170. Wolfgang Kruse, "Der säkularisierte 'heilige Krieg' des deutschen Reiches 1914," *Journal Geschichte*, May 1989, 23; Mosse, 33.

171. Theodor Heuss, *Errinerungen, 1905–1933* (Tübingen: Leins, 1963), 140.

172. Rolf Marben, *Ritter der Luft* (Hamburg: Broscher, 1931), 3.

Chapter Four. The Airship as a Business Tool in Weimar Culture

1. Laszlo Moholy-Nagy, "Constructivism and the Proletariat" (1922), reprinted in Richard Kostelanetz, ed., *Moholy-Nagy* (New York: Praeger, 1970), 185. I am grateful to Archie Newton Perrin for this reference. Detlev Peukert, *The Weimar Republic: The Crisis of Classical Modernity* (New York: Hill & Wang, 1992), 163–67; Jost Hermand, "Unity within Diversity? The History of the Concept of 'Neue Sachlichkeit,'" in *Culture and Society in the Weimar Republic*, ed. Keith Bullivant (Manchester: Manchester University Press, 1977), 166–67.

2. Alf Lüdtke, "Ikonen des Fortschritts," in *Amerikanisierung: Traum und Alptraum im Deutschland des 20. Jahrhunderts* (Stuttgart: Franz Steiner, 1996), 201.

3. Alfred Colsman, *Luftschiff voraus!* (1933; reprint, Munich: Zuerl, 1983), 207–27.

4. Wolfgang Meighörner, *Alfred Graf von Soden-Fraunhofen* (Friedrichshafen: Zeppelin Museum, 1994), 39–45.

5. Peter Kleinheins, *LZ 120 "Bodensee" and LZ 121 "Nordstern"* (Friedrichshafen: Zeppelin Museum, 1994), 26–27.

6. Ibid., 55, 64.

7. NASM, A3G-309320-01, Col. Hensley, "Report on the Bodensee German Airship."

8. PA, R32557, R32558, R32559; Kleinheins, *LZ 120,* 96–127.

9. PA, R32557, AA memorandum on Swiss minister de Planta's inquiries regarding civilian airship operations, 11 May 1920.

10. PA, R32558, protocol, 30 June 1921.

11. PA, R32807, letter of Major Tanner, Munich, to AA, 8 February 1921; *Kreuzzeitung,* 3 October 1920. Newspaper commentaries included the suggestion that the aviation sheds be left standing as a way of relieving a dreadful housing shortage further aggravated by unemployment. A Swiss commentator sympathetic to the German position suggested that Allied calls for the complete destruction of the airship works amounted to "spiritual murder."

12. PA, R32929, letters of German embassy in Madrid to AA, 16 July and 2 August 1921; *London Times,* 27 September 1921. The interest in building a permanent airship port in Seville, Spain, remained a fixture of Zeppelin politics until the beginning of the Spanish Civil War.

13. Richard K. Smith, *The Airships Akron and Macon* (Annapolis, Md.: Naval Institute Press, 1965); Douglas Robinson and Charles Keller, *Up Ship! U.S. Navy Rigids, 1919–1935* (Annapolis, Md.: Naval Institute Press, 1982); William F. Althoff, *Skyships* (New York: Orion Books, 1990).

14. NASM, A3G-3090030-01, memorandum of the American mission in Berlin, 12 September 1920; A3G-309040-01, memorandum of 27 October 1921; Robinson and Keller, chaps. 1–2; cf. Henry Cord Meyer, *Airshipmen, Businessmen, and Politics, 1890–1940* (Washington, D.C.: Smithsonian, 1991), chap. 5.

15. PA, R32558, letter of Mastermann to the peace section *(Friedensabteilung)* of the AA, 28 February 1922; *Vossische Zeitung,* 4 March 1922; *Chicago Tribune,* 16 March 1922.

16. PA, R32558, letter of Zeppelin company to AA, 28 February 1922.

17. PA, R32558, AEG credentials letter to AA, 21 June 1922. Other airship reparations also followed, such as the transfer of an airship hangar to Japan. The American agreement included the requirement that Zeppelin be able to furnish financial and industrial assets in excess of 3 million gold marks—the estimated value of the new airship—in case delivery of the dirigible could not take place.

18. PA, R32558, agreements A and B (English and German) on the application of Ambassadors' Conference resolution of 157th meeting, 16 December 1921; *Vossische Zeitung,* 14 October 1924; cf. *Zeppelin Post Rundbrief* 1/1991, 4. Some

sources suggest that Rathenau, who was assassinated on 24 June 1922, was on his way to signing the agreement with U.S. ambassador Houghton, and that Secretary of State Haniel von Haimhausen signed on his behalf. Although the latter point is correct, the agreement is dated 26 June, and it remains unclear whether Rathenau was indeed supposed to have signed it.

19. PA, R32558, letter of Zeppelin company to RFM, 16 March 1923.

20. PA, R32559; NASM, A3G-3090030-01; Meyer, 53–80, 128–30; cf. Dorothea Haaland, *Der Luftschiffbau Schütte-Lanz Manheim-Rheinau (1909–1925): Die Geschichte einer innovativen Idee als zeitlich räumlicher Prozess* (Mannheim: Institut für Landeskunde und Regionalforschung, 1987).

21. BA, R43II/698, 46–47, letters of Colsman to AA, 3 March 1921, and to Reich Chancery (State Secretary Albert), 4 March 1921; cf. Meyer, 136; Lutz Budrass, *Flugzeugindustrie und Luftrüstung in Deutschland, 1918–1945* (Düsseldorf: Droste, 1998), 97–98.

22. BA, NL96/7 (Dörr), Freiherr von Gemmingen, undated memorandum [1918–19].

23. BA, R43II/698, 61, letter of State Secretary von Haniel to Zeppelin company, 21 March 1921.

24. PA, R32559, AA memorandum, 25 January 1924; Robinson and Keller, 177–88; Meyer, chaps. 5 and 9; Hans G. Knäusel, *Zeppelin and the United States of America: An Important Episode in German-American Relations*, 2d ed. (Friedrichshafen: Zeppelin, 1981), chap. 2.

25. BDM, Poeschel papers, unattributed clipping, 3 August 1921; Douglas Robinson, *Giants in the Sky: A History of the Rigid Airship* (Seattle: University of Washington Press, 1973), 162–67.

26. PA, R32559, *Berliner Montagspost,* 17 March 1924.

27. BDM, Poeschel papers, letter of Willy Fisch (RVM official) to Johannes Poeschel, 9 October 1924.

28. PA, R32559, letter of Colsman to AA, 14 May 1924; *Vorwärts,* 3 October 1924; cf. *Vossische Zeitung,* 16 October 1924; PA, R32558, *Berliner Tagblatt,* 2 October 1924 (evening ed.); *Tag,* 29 September and 3 October 1924.

29. Ludwig Dürr, *Fünfundzwanzig Jahre Zeppelin-Luftschiffbau* (Berlin: VDI, 1925), reprinted in Peter Kleinheins, *Die grossen Zeppeline* (Düsseldorf: VDI, 1985), 16–19.

30. Ibid., 43.

31. *Kölnische Zeitung,* 25 September 1923, trans. in *The Living Age,* 17 November 1923, 318.

32. Dürr, 94–96.

33. *Vorwärts,* 14 October 1924.

34. *Vossische Zeitung,* 8 September 1924 (evening ed.); Gustav Radbruch, *Biographische Schriften,* ed. Günter Spendel (Heidelberg: C. F. Müller, 1988), 62.

35. *Vorwärts,* 11 and 13 October 1924.

36. *Die Rote Fahne,* 15 October 1924.

37. Meyer, chap. 5.

38. *Berliner Tagblatt,* 2 October 1924.

39. Artist E. Schilling drew "Der Amerikafahrer."

40. Johannes Baader Oberdada, *Das Geheimnis des ZR III* (Berlin: Stiemer, 1924), reprinted in *Reisen im Luftmeer,* ed. Karl Riha (Munich: Hanser, 1983), 282–90.

41. WTB, 15 October 1924.

42. LCM, Hildebrandt papers, box 17, *Der Schild* ("Zeitschrift des Reichsbundes jüdischer Front Soldaten e. V., Berlin") 15 February 1925.

43. *Deutsche Zeitung,* 13 October 1924.

44. *Vossische Zeitung,* 15 October 1924; Harold Dick and Douglas Robinson, *The Golden Age of the Great Passenger Airships Graf Zeppelin and Hindenburg* (Washington, D.C.: Smithsonian, 1985), 173. Arnstein (1887–1974), said to have suffered from pervading antisemitism in Friedrichshafen too, emigrated to America in 1924 to work for the Zeppelin-Goodyear joint venture.

45. Fritz Müller, "Die Amerikafahrt des Zeppelin im Lichte des Rechts," *Deutsche Juristen-Zeitung* 21–22 (1 November 1924).

46. PA, R32959, AA memorandum, 25 October 1924.

47. Hans von Schiller, *Zeppelin: Aufbruch ins 20. Jahrhundert,* 2d ed., ed. Hans G. Knäusel (Bonn: Kirschbaum, 1988), 210–11. The maximum permissible airship size under the Versailles treaty was a volume of 30,000 cubic meters, which was roughly the size of a 1915–built airship. End-of-the-war airships were double that size. ZR 3 was some 70,000 cubic meters in volume.

48. PA, R32808, letter of RVM to AA, 22 September 1924.

49. *Der Tag,* 28 September 1924, transcription of Gustav Stresemann address to a DVP party gathering (27 September 1924).

50. PA, R32809, letter of the Frankfurt League for Aviation to Chancellor Marx, 12 January 1925; cf. letter of the Hildesheim Association for Aviation to the "Reich Presidents" *(sic),* 15 January 1925; cf. Cassel and Halle branches of the Reich Union of German Technology (Reichsbund deutscher Technik), 6 January and 13 February 1925; Bad-Pfalz Aviation Association, 25 January 1925.

51. PA, R32808, letter of the Reichsarbeitgemeinschaft technischer Beamtenverbände (Reich Association of Technicians) to Reich Chancery, 28 October 1924; Reichstag query 68 (16 October 1924). The Pomeranian section of the German Aviation Association (DLV) issued its plea on the eighth anniversary of the death of war ace Oswald Boelcke (28 October 1924).

52. PA, R32808, *Hannoverscher Kurier,* 3 November 1924; cf. *Basler Nachrichten,* 17 October 1924.

53. PA, R32808, letter of AA to RVM, 11 October 1924; *Vossische Zeitung,* 14 October 1924; cable of Paris embassy to AA, 22 October 1924.

54. PA, R32808, cables of AA to the German embassy in Washington, October 1924; cable of AA to Paris embassy, 21 October 1924; cables of London embassy to AA, 22 and 23 October 1924; minutes of meeting between RVM and AA, 23 October 1924. William Randolf Hearst's *New York American* ran a column advocating, "Let them build more Zeppelins," arguing that Germany would thus increase exports and be able to repay loans. Lord Thompson had promised he would raise the issue with Prime Minister MacDonald as well as French air minister Eynac and Marshall Foch.

55. PA, R32559, letter of Oslo embassy to AA, 16 October 1924.

56. *Kölnische Zeitung*, 25 September 1923, trans. in *The Living Age*, 17 November 1923, 318; *Kölnische Zeitung*, 6 November 1924; cf. *Arbeiter Zeitung* (Vienna), 16 October 1924.

57. PA, R32808, *8 Uhr Abendblatt*, 23 October 1924; PA, R32985, *Frankfurter Kurier*, 7 September 1924.

58. PA, R32808, letter of The Hague embassy to AA, 20 October 1924; *Neue Zürcher Zeitung*, 16 October 1924; *Tages Anzeiger* (Zurich), 13 September 1924.

59. *Neue Freie Presse*, 16 and 17 October 1924; PA, R32808, Vienna embassy to AA, 17 October 1924.

60. John Hiden, "The Weimar Republic and the Problem of the Auslandsdeutsche," *Journal of Contemporary History* 12 (1977): 273–89.

61. PA, R32809, *Volksbund "Rettet die Ehre,"* brochure to the Government of the Republic of Ecuador, undated [1924]. The chairman of the league, O. Hartwich, was described as a "cathedral preacher."

62. Hiden, 279.

63. PA, R32559, letter of German consulate in Abyssinia to AA, December 1924; letter of Lima embassy to AA, 2 April 1925.

64. PA, Presse Abt. Fach 125, Regal 26, Packet 256, Deutschland 9, vol. 1; letter of *Chicago Abendpost* correspondent to AA, 27 September 1924; Reich President's Office memorandum, 29 September 1924.

65. Robinson, *Giants*, 259–60.

66. PA, R32809, Vienna embassy to AA, 2 March 1925.

67. Hugo Eckener, *Im Zeppelin über Länder und Meere* (Flensburg: Christian Wolff, 1949), 98; cf. Robinson, *Giants*, 261.

68. Wolfgang Meighörner, "25 Jahre Zeppelin-Luftschiffbau: Eine Gedenkfeier für die Zukunft," in *Zirkel, Zangen, und Cellon: Arbeit am Luftschiff*, ed. Wolfgang Meighörner (Friedrichshafen: Zeppelin Museum, 1999), 109.

69. BA, R43I/737, 9, telegram of Chancellor Luther to Eckener, 20 August 1925.

70. BA, R43I/737, description of the *Spende*, 4 August 1925.

71. *Luftfahrt*, 20 July 1926.

72. Russell A. Berman, *Cultural Studies of Modern Germany: History, Representation, and Nationhood* (Madison: University of Wisconsin Press, 1993), 26–27.

73. BA, R43I/737, memorandum of the State Commissioner for Welfare Services, 23 July 1925.

74. BA, R43I/737, circular from the Office of the Minister-President of Prussia, 5 August 1925; Peter Fritzsche, *A Nation of Fliers: German Aviation and the Popular Imagination* (Cambridge: Harvard University Press, 1992), 142.

75. BA, R43I/737, protocol of ministerial meeting, 10 August 1925.

76. BA, R43I/737, letter of Eckener to the Reich chancellor, 15 August 1925.

77. BA, R43I/737, Chancery memorandum, 15 September 1925. RVM and AA concurred in their evaluations of the *Spende*'s foreign political implications.

78. BA, R43I/737, minutes of meeting between Eckener and governmental representative, 16 September 1925.

79. GStA, Rep120 Tn35, summary of fund-raising to 31 December 1926. Of the almost 2.6 million marks raised by that time, approximately 17,500 marks came from abroad.

80. BA, R43I/737, letter of RWM to Chancery, 22 March 1926.

81. HAStK, Best 902 (Adenauer), 233, Fasz. 1, 175–76, telegram from the *Spende* committee to Adenauer, 12 October 1925; HHW, 405/I-P/4, vol. 5542, *Wiesbadener Tageblatt*, 17 October 1925. Adenauer's response is not known. At the time, however, he was heavily involved in all aviation matters as a member of a special board that helped pave the way for the founding of Lufthansa that same year.

82. BA, R43I/737, Reich Chancery report on Prussian government ministerial meeting, 16 September 1925; cf. WTB, 22 September 1925.

83. BA, R43I/737, excerpts from ministerial meeting minutes, 21 September 1925.

84. BA, R43I/737, letter of Eckener to Reich President's Office, 21 April 1926; cf. Fritzsche, 142.

85. Hans Hildebrandt, ed., *Zeppelin—Denkmal für das deutsche Volk* (Stuttgart: Germania, 1925). A second edition of this book, published in 1931, had a different editor, and several of its original contributions were withdrawn and others added.

86. Hermann von Müller, *Herrscher im Reiche der Technik* (Leipzig: Teubner, 1931), 45–59; cf. Heinrich Netz, *Berühmte Männer der Technik* (Leipzig: Quelle & Meyer, 1930), 126–47.

87. GStA, Rep120 Tn35, vol. 1, correspondence concerning *Spende* issue.

88. BA, NL96/11 (Dörr), "Der neue Zeppelin," n.d. [1927].

89. *Straubinger Tagblatt*, 2 October 1927.

90. Fritzsche, 143.

91. Eckener, 104; correspondence of the author with Henry Cord Meyer, 13 June 1995.

92. *Deutsche Allgemeine Zeitung*, 18 February 1928; cf. PA, R32986, *Kreuzzeitung*, 13 February 1928; *Berliner Börsen Zeitung*, 14 February 1928.

93. *Kölnische Illustrierte Zeitung*, 29 September 1928, 1215; *Schlesische Zeitung*, 17 February 1928.

94. BA, NL96/14 (Dörr), *Straubinger Tagblatt*, 2 October 1927.

95. Dürr, 159; Dick and Robinson, 34.

96. PA, R32986, *Lokal Anzeiger*, 21 September 1928; *Berliner Tageblatt*, 3 October 1928; *Berliner Börsen Zeitung*, 3 October 1928. There is no evidence suggesting that the airship intentionally flew over the Rhineland. The German press also cited Allied reactions to the flights.

97. *Kölnische Illustrierte Zeitung*, 29 September 1928.

98. BA, Stresemann papers, roll FC 483 (vol. 72), telegram to Eckener, 22 September 1928; cf. Eckener, 94–97.

99. Ludwig Dettmann, *Mit dem "Zeppelin" nach Amerika* (Berlin: Reimar Hobbing, 1929).

100. *Der Abend* (evening supplement to *Vorwärts*), 3 October 1928; *Schlesische Zeitung*, 28 September 1928.

101. Lynn Abrams, "From Control to Commercialization: The Triumph of Mass Entertainment in Germany, 1900–1925," *German History* 8 (1990): 278–93; cf. Karl Christian Führer, "A Medium of Modernity? Broadcasting in Weimar Germany, 1923–1932," *Journal of Modern History* 69 (December 1997): 722–53.

102. *Rote Fahne*, 3 and 19 October 1928.

103. Willy Reese, *"Z.L. 127": Das fliegende Hotel* (Leipzig: Deutsche Buchwerkstätten, 1928).

104. Georg Paul Neumann, *In der Luft unbesiegt* (Munich: Lehmanns, 1923); Waldemar Koelle, "Marineluftschiffe im Kriege, in Sturm und Not: Errinerungen," *Deutsche Rundschau* 205 (December 1925): 243–62; *Kölnische Illustrierte Zeitung*, 23 June 1928, 784–85; Michael Gollbach, *Die Wiederkehr des Weltkrieges in der Literatur: Zu den Frontromanen der späten zwanziger Jahre* (Kronberg T.s.: Scriptor, 1978); cf. George L. Mosse, *Fallen Soldiers: Reshaping the Memory of the World Wars* (New York: Oxford University Press, 1990).

105. Robinson, *Giants*, 97–102.

106. *Kölnische Illustrierte Zeitung*, 3 January 1931, 17–19; LCM, Hildebrandt papers, box 14, *Allensteiner Volksblatt*, 25 August 1930. As if to confirm the place of the airship in the war heritage, the *Graf Zeppelin* was flown to the Tannenberg monument on the occasion of the unveiling of the plaque in memory of German World War I fliers.

107. Horst von Buttlar Brandenfels, *Zeppeline gegen England* (Zurich, Leipzig: Almathea, 1931), 6.

108. Ibid., 11; Fritz Strahlmann, ed., *Zwei deutsche Luftschiffhäfen des Weltkrieges: Alhorn and Wildeshausen* (Oldenburg: Lindenallee, 1926), introduction; Rolf Marben, *Ritter der Luft: Zeppelinabenteuer im Weltkrieg* (Hamburg: Broschek, 1931), introduction.

109. Brandenfels, 208.

110. Fritzsche, 146.

111. *Arbeiter Zeitung* (Vienna), 17 October 1928.

112. *Illustrierte Zeitung*, 17 November 1928.

113. Jonathan Woodham, *Twentieth Century Ornament* (New York: Rizzoli, 1990), 116–22.

114. Claude Lichtenstein, ed., *Streamlined: A Metaphor for Progress* (Baden, Switzerland: Lars Müller, n.d.), 9–13.

115. Roger Bilstein, "Air Travel and the Traveling Public: The American Experience, 1920–1970," in *From Airships to Airbus: The History of Civil and Commercial Aviation*, vol. 2, ed. William F. Trimble (Washington, D.C.: Smithsonian, 1995), 94. Unpleasant experiences were the lot of airline passengers throughout the world.

116. Heinrich Eduard Jacob, *Mit dem Zeppelin nach Pernambuco: Poetische Luftbilder einer ungewöhnlichen Reise* (Berlin: Katzengraben Presse, 1992), 13.

117. Ibid., 27.

118. *Illustrierte Zeitung*, 13 and 27 October 1928, 30 November 1929.

119. Michael Neufeld, "Weimar Culture and Futuristic Technology: The Rocketry and Spaceflight Fad, 1923–1933," *Technology and Culture* 31, 4 (October 1990): 725–52; Adelheid von Saldern, "Cultural Conflicts, Popular Mass Culture, and the Question of Nazi Success: The Eilenriede Motorcycle Races, 1924–1939," *German Studies Review* 15, 2 (May 1992): 317–38; Sigfried von Weiher, "Franz Kruckenbergs Lebenswerk, 1882–1982," *Kultur und Technik* 6, 4 (December 1982): 226–30. The engineer who designed the train had actually worked in the past for Schütte-Lanz, Zeppelin's competitor, but the "rail Zeppelin" nickname stuck.

120. Ernst Baum, ed., *Frohes Schaffen: Das Buch für jung und alt* (Vienna, Leipzig: Deutscher Verlag für Jugend und Volk, 1929), 59–78.

121. Hartwig Fritzsche school trip recollection, *Leben und Arbeit* (Hermann-Litz Schule) 2/3 (1930–31): 87–88.

122. Erika Mann, *Stoffel fliegt übers Meer* (1932); the title page picture, representing a hangared airship, is reproduced in Otto Brunken et al., *Geschichte der deutschen Kinder- und Jungenliteratur* (Stuttgart: Metzler, 1990), 293; Robert Theuermeister, *Vom Luftballon zum Zeppelin* (Leipzig: Ernst Wunderlich, 1931), introduction.

123. Joachim Breithaupt, *Mit Graf Zeppelin nach Süd- und Nordamerika: Reiseeindrücke und Fahrterlebnisse* (Lahr [Baden]: Moriss Schauenburg, 1930), introduction.

124. Cf. von Saldern, 324–25.

125. *Kölnische Zeitung*, 29 March 1929; *Deutsche Allgemeine Zeitung*, 31 March 1929.

126. PA, R32992, AA memorandum, 17 June 1930; letter of Paris embassy to AA, 19 July 1930. The German ambassador in Paris suggested that while sale of the coins in France was a bad idea, offering them as a gift to selected French officials

might foster goodwill and smooth the way for the granting of overflight permission to the airship; cf. PA, R32992, requests of the "Deutsch-Asiatische" Bank, Shanghai, 28 June 1930.

127. Max Geisenheyner, *Mit "Graf Zeppelin" um die Welt* (Frankfurt a.M.: n.p., 1929), 59–64.

128. Jerónimo Megías, *La primera vuelta al mundo en el "Graf Zeppelin"* (Madrid: Hauser & Menet, 1930); Léo Gerville Réache, *Autour du monde en Zeppelin* (Paris: Nouvelle Revue critique, 1929).

129. Ludwig Fischer, ed., *Graf Zeppelin—Sein Leben—Sein Werk* (Munich: Oldenbourg, 1931), 306.

130. BA, R43I/737, 176. Brandenburg held the rank of *Ministerialdirigent* (three levels down from minister) and had thus bypassed several people's authority in calling for the official flagging.

131. BA, R43I/737, 176–77.

132. *General Anzeiger* (Bonn), 23 April 1930.

133. *General Anzeiger* (Bonn), 22 and 23 April 1930; cf. SWB, Pr82/16, *Bochumer Anzeiger,* 24 April 1930; *Kölnische Zeitung,* 23 April 1930. Newspapers outside of Bonn chose to emphasize the lack of mass organization.

134. *Kölnische Illustrierte Zeitung,* 12 July 1930, cover.

135. *Kölnische Zeitung,* 7 July 1930. Cologne mayor Konrad Adenauer, who had lobbied hard for the show to take place, oversaw much of the preparations.

136. *Kölnische Zeitung,* speech of Transportation Minister von Guérard, 7 July 1930.

137. Stanley Suval, *The Anschluss Question in the Weimar Era* (Baltimore, Md.: Johns Hopkins University Press, 1974), chaps. 1 and 2.

138. Ibid., 194.

139. PA, R32989, letter of Austrian transport minister Schürff to Ambassador Lerchenfeld-Köfering, 16 May 1929.

140. *Berliner Lokalanzeiger,* 29 March 1929.

141. *Germania,* 3 April 1929.

142. SWA, F25/149, correspondence pertaining to Schwarz controversy (1931); letter of TMW to Carl Berg, December 1930. The Technology Museum in Vienna had plans to include an exhibit on Schwarz as part of an Austrian initiative. Berg published a pamphlet entitled *David Schwarz—Carl Berg—Graf Zeppelin: Ein Beitrag zur Entstehung der Zeppelin-Luftschiffahrt* (1926), which presented various aspects of the evolution of the rigid-airship concept.

143. AdR, 10/4, IIIGf, 39059/29, box 204.

144. *Neue Freie Presse,* 5, 12, and 13 July 1931; *Illustrierte Kronen Zeitung,* 12 July 1931.

145. *Der Tag* (Vienna), 12 July 1931.

146. *Flug* (Vienna), June–July 1931; *Die Rote Fahne* (Vienna), 12 and 14 July

1931. The *Rote Fahne* was the only newspaper to criticize the pouring of 60,000 schillings into the airship rather than into repairs to a Vienna bridge. It also reported on the death of a youth who had fallen after climbing onto a roof to watch the airship over Vienna.

147. *Kölnische Illustrierte Zeitung*, 31 August 1929, 1098. The article described Eckener as the upholder of traditions set by Magellan, Francis Drake, and James Cook.

148. *New York Times*, 12 April 1930.

149. Meyer, 261; correspondence of the author with Henry Cord Meyer, 13 June 1995.

150. BA, NL96/14 (Dörr), *Straubinger Tagblatt*, 2 October 1927.

151. PA, R32990, letter of Gustav Adolf Langen to Eckener, 10 December 1929, transcribed in letter of Eckener to AA, 17 December 1929.

152. PA, R32992, letter of Washington embassy to AA, 25 June 1930; cf. MHS, Hammond James Dugan papers (ms. 1859), box 1, letter to his wife, 5 November 1928. Dugan was on hand during the first visit of the *Graf* and, commenting on the crowds, described the experience as an unforgettable event that should never be repeated.

153. PA, R32992, letter of Merchants' Association of New York to German-American Chamber of Commerce, 29 May 1930.

154. PA, R32992, letter of New York consulate to AA, 12 April 1930; cf. *New York Times*, 12 April 1930. Eckener made these remarks during an impromptu radio address on the occasion of a dinner given in New York by the Association of Former German Students in America.

155. Fritzsche, 176.

156. Ibid., 175.

157. LCM, Hildebrandt papers, box 17, *Der Reichsbote*, 17 December 1932; cf. *Magdeburgische Zeitung*, 17 December 1932; *Weserzeitung*, 19 December 1932; cf. unattributed clipping, 18 May 1932; cf. *Berliner-Börsen Zeitung*, 8 July 1932.

158. BA, R2/5607, correspondence regarding funding matters.

159. Bernard von Brentano, *Wo in Europa ist Berlin?* (Frankfurt a.M.: Surkamp, 1987), 103–8; cf. Fritzsche, 135–37.

160. SWA, K3/6550, "Planung für Zeppelin Landung in Bielefeld," n.d.

161. SWA, K3/6550, Bielefeld memorandum, July 1930.

162. R. Lapatki and K. Renn, *Vom Schwabemeer . . . zur Theaterstadt im Luftschiff* (Meiningen: Philatelistenverband, 1981), 16–19.

163. Peukert, 166; David Nye, *American Technological Sublime* (Cambridge: MIT Press, 1994), 202–3.

164. *Plakatsammlung des Instituts für Zeitungsgeschichte, Dortmund; Deutsche Plakate*, part 4, *Weimar Republic* (Munich: K. G. Saur, n.d.), microfiche 70, plate 32. The undated poster [1930 or 1932] was produced by "Fritz Arbeit" and shows

a Zeppelin flying over the Reichstag and dropping leaflets recommending the Deutsch-National Volskpartei as the choice for all voters.

165. BA, R43I/586, letter of Eckener to Brüning, 18 March 1932, and response, 4 April 1932. The broadcast speech was given at a rally in the Sportpalast in Berlin on 8 April.

166. Arnold Brecht, *Mit der Kraft des Geistes: Lebenserrinerungen,* vol. 2 (Stuttgart: Deutsches Verlangsanstalt, 1967), 152–53.

167. Hans Fass, "Die Führung der Technik," *Technik und Kultur* 16, 4 (15 April 1925): 53; Jeffrey Herf, *Reactionary Modernism: Technology, Culture, and Politics in Weimar and the Third Reich* (Berkeley: University of California Press, 1985), 1–17.

168. Peukert, 168.

169. Gustav Radbruch, *Politische Schriften aus der Weimarer Zeit—1. Demokratie, Sozialdemokratie, Justiz,* ed. Alessandro Baratta (Heidelberg: C. F. Müller, 1992), 62.

170. *Vossische Zeitung,* 10 September 1924.

171. LCM, Hildebrandt papers, box 16, letter of Colsman to Hildebrandt, 17 October 1928.

172. Herf, 226. The typology lists business *(Wirtschaft)* as a direct opposite of *Kultur.*

173. Peter S. Fisher, *Fantasy and Politics: Visions of the Future in the Weimar Republic* (Madison: University of Wisconsin Press, 1991), 104–56; Stephen Lamb and Anthony Phelan, "Weimar: The Birth of Modernism," in *German Cultural Studies: An Introduction,* ed. Rob Burns (New York: Oxford University Press, 1995), 70–71.

Chapter Five. Ideologies of Science and Adventure: The Arctic Airship

1. Raoul Girardet, *Mythes et mythologies politiques* (Paris: Seuil, 1986), 13.

2. David H. DeVorkin, *Race to the Stratosphere: Manned Scientific Ballooning in America* (New York: Springer, 1989), 3.

3. As early as 1709 the Portuguese monk Gusmão predicted an exploration of the North Pole by means of an airship. Starting in 1845 many enthusiasts published proposals for such a quest.

4. S. A. Andrée et al., *Andrée's Story: The Complete Record of His Polar Flight, 1897,* ed. Swedish Society for Anthropology and Geography (1930; reprint, New York: Viking, 1958).

5. *Wiener Luftschiffer Zeitung,* May 1906, 103; "S. A. Andrée, ein Schicksalsgefährte Nobiles," *Luftfahrt,* 22 June 1928, 177–79.

6. John Grierson, *Challenges to the Poles* (Hamdon: Archon Books, 1964), 23; cf. George Simmons, *Target Arctic: Men in the Skies at the Top of the World*

(Philadelphia: Chilton, 1965), appendix 2; *Wiener Luftschiffer Zeitung*, August 1905, 168. Among Andrée's heirs, the Frenchman Marcillac designed a motorized balloon; meanwhile, the American Walter Wellman made two attempts to fly to the Pole over the period 1906–10.

7. P. J. Capelotti, *The Wellman Polar Airship Expeditions at Vigohmna, Danskøya, Svalbard* (Oslo: Norsk Polarinstitutt, 1997), 85.

8. O. Bashin, "Berichterstattung aus der Arktis," *Zeitschrift für Geopolitik* 1, 8 (1924): 511–17; Michael Adas, *Machines as the Measure of Men: Science, Technology, and Ideologies of Western Dominance* (Ithaca, N.Y.: Cornell University Press, 1989), chap. 4.

9. Frank Petsch, "Scientific Organization and Science Policy in Imperial Germany, 1871–1914: The Foundation of the Imperial Institute of Physics and Technology," *Minerva* 7, 4 (October 1970): 558.

10. Ibid., 559–61.

11. Hans-Christian Täubrich and Jutta Tschoeke, eds., *Unter Null: Kunsteis, Kälte, und Kultur* (Munich: Beck, 1991).

12. A. Wellner drawings in *Lustige Blätter* (1909), reprinted in *Kölnische Illustrierte Zeitung*, 24 August 1929, 1094.

13. Dr. Elias, "Die Erreichung der Pole," in *Wir Luftschiffer*, ed. Karl Bröckelmann (Berlin: Ullstein, 1909), 416–33. An illustrated album, *Mit Zeppelin zum Pol!* (Neurode: n.p., 1909), also anticipated an airship expedition to the region.

14. LZA, 03/016; BA, R5/3833, report 15492 of the Interior Ministry of Württemberg, 3 September 1909. The Zeppelin Bund was established by the Stuttgart librarian Fritz Lehmann in July 1909. Open to all for a modest contribution, it was to keep its supporters informed through its *Augen auf* periodical, which covered both aeronautical matters and more general cultural topics ("Freie Vereinigung für Naturfreude und Wanderlust"). The association also worked toward the establishment of a Zeppelin museum. In late August, however, Count Zeppelin withdrew his support, and the association had to change its name.

15. *Vossische Zeitung*, 10 September 1924; H. Bausinger et al., eds., *Reisekultur* (Munich: Beck, 1991), 249–61. Ernst Arnd published several journals based on his trips to Scandinavia and Russia.

16. Carl Waack, *Von Andree bis Zeppelin: Das Luftschiff im Dienste der Polarforschung* (Rostock: Volckmann, 1910), 9.

17. Rudolf Martin, *Das Zeitalter der Motorluftschiffahrt* (Leipzig: Theod. Thomas, 1907), 97.

18. The count may have also become interested in the region through his cousin Max, who had participated in an 1891 expedition to the Arctic Circle.

19. Rolf Italiaander, *Ferdinand Graf von Zeppelin* (Konstanz: Stadler, 1986), 131; *Deutsche Zeitschrift für Luftschiffahrt*, 29 June 1910, 16–17; cf. *Globus* 97, 16 (28 April 1910): 251–53; BA, R5/3847, letter of Zeppelin to Chancellor's Office,

9 December 1909; Cornelia Lüdecke, *Die deutsche Polarforschung seit der Jahrhundertwende und der Einfluss Erich von Drygalskis* (Bremerhaven: Alfred-Wegener-Institut, 1995), 52 n. In 1910 a controversy arose from a suit brought against Zeppelin by the explorer Theodor Lerner, who claimed that he had suggested the polar airship idea to the count and was thus entitled to a seat on the Zeppelin Polar Committee.

20. *Der Tag*, 2 July 1909. Hugo Hergesell (1859–1938) was very active in aeronautical circles, promoting the study of aviation meteorology from his home in Strasbourg. He and Zeppelin collaborated early on when the former set up a kite station on the *Bodensee* to study wind conditions.

21. BA, R5/3847, letter of German embassy in Christiania to AA, 15 October 1909.

22. LZA, 03/005, letter to Count Zeppelin, 24 March 1910, with letter of Heineken, director of the Norddeutsche Lloyd, enclosed.

23. BA, R5/3847, letter of Treasury Office to Chancellor's Office, 7 March 1910.

24. *Petermanns Geographische Mitteilungen*, 4 and 5 (1911): 178–80, 241–46; cf. BDM, 1949-151, "Zeppelin-Expedition nach Spitzbergen."

25. A. Miethe and H. Hergesell, *Mit Zeppelin nach Spitzbergen* (Berlin, Vienna: Deutsches Verlaghaus Bong & Co., 1911), 1–2.

26. H. Hergesell, *Unsere geplante Polarfahrt mit Zeppelin Schiffen*, series Petermanns Mitteilungen no. 57 (Gotha: Justus Perthes, 1911); LZA, 03/005, Hamburger Verein für Luftschiffahrt (Hamburg Aeronautical Association) to Hergesell, 14 March 1910; A. Stolberg, "Über die Verwendung lenkbarer Ballone im Hohen Norden," *Deutsche Rundschau* 150 (January–March 1912): 464–69.

27. LZA, 04/197; LZA, 04/248, telegram from Spitzbergen expedition to Count Zeppelin, 21 April 1914; LZA, 03/01.

28. LZA, 04/248.

29. *Deutsche Luftschiffer Zeitung*, 22 December 1918, 7–8. The only reported flights to the Arctic Circle during the war took place in 1914–15 out of Russia.

30. *Nachrichten für Luftfahrer*, 26 February 1922, 103; cf. *Der Luftweg*, 12 November 1922, 137.

31. PA, R32458, letter of AA to Stockholm embassy, March 1925; Leonid Breitfuss, *Aeroarctic*, series Pettermans Mitteilungen no. 201 (Gotha: Justus Perthes, 1927), 14. Bruns contacted the Zeppelin company twice, on 19 August 1920 and 19 April 1922, regarding Arctic plans. Eckener answered the second letter and declined, for reasons of work, to help Bruns publicize his project.

32. *Nachrichten für Luftfahrer*, 26 February 1922, 103; cf. *Münchener Neueste Nachrichten*, 28 March 1922.

33. BA, NL96/8 (Dörr); cf. letter of Hans von Schiller to *Buoyant Flight* 17, 5 (March 1970). The navy Zeppelin L 59 successfully flew from Bulgaria to North Africa in 1917. As for Zeppelin L 72, it evidently had enough autonomy to reach

New York. The successful Atlantic crossing by the British R 34 airship in 1919 does not appear to have had any particular influence on expert opinions.

34. Steinar Kjaerheim, ed., *Fridtjof Nansen Brev IV (1919–1925)* (Oslo: Universitetsforlaget, 1966), letters of Nansen to Bruns, 25 March 1922, and to Kohlschütter, 2 August 1924; Detlef Brennecke, *Fridtjof Nansen* (Hamburg: Rowohlt, 1990), 124.

35. *Das Luftschiff als Forschungsmittel in der Arktis: Eine Denkschrift* (n.p.: Aeroarctic, 7 October 1924); cf. BA, R43II/698, 291, letter of Aeroarctic to Chancery, 28 December 1925; Henry Cord Meyer, *Airshipmen, Businessmen, and Politics, 1890–1940* (Washington, D.C.: Smithsonian, 1991), 159; cf. Lüdecke, 163. There is much confusion and contradiction in both primary and secondary sources as to the date of Aeroarctic's founding and the identity of those involved.

36. Paul Forman, "Scientific Internationalism and the Weimar Physicists," *Isis* 64 (1973): 153–55.

37. PA, R32804, letter of Stockholm embassy to AA, 13 November 1924; Meyer, 156–57. The ambassador warned Berlin not to make any comments on the "German-friendly" action, since many academy members were "ententophiles."

38. Forman, 161.

39. PA, R32804, letters of Stockholm embassy to AA, 18, 25, 31 October, and 11 November 1924; cf. Meyer, 156. Hedin had met Count Zeppelin before the Great War; it is not known whether they discussed the matter of exploration via airship.

40. PA, R32458, letter of Stockholm embassy to AA, 29 November 1924; cf. Meyer, 159. Bruns estimated that total costs for the airship and the expedition would reach some 8.5 million gold marks. Hedin hoped also to use the machine for a trip to Tibet.

41. PA, R32458, letter of AA to Eckener, 4 December 1924. Proceeding on the assumption that few nations could afford such an investment, Privy Counselor Nord suggested that aside from England, which wanted to link London to Egypt and India, only the United States with its plans for an Alaskan base was a true contender. The Soviet Union did not come into consideration until early 1925. Meyer, 160; Hugo Eckener, *Im Zeppelin über Länder und Meere* (Flensburg: Christian Wolff, 1949), 357. Eckener claimed in his memoirs that independent German newspapers had feared opposing the nationalist "trumpets" organized by an unnamed German press tycoon (probably Alfred Hugenberg) in opposition to any international cooperation over the airship.

42. PA, R32458, letter of Eckener to AA, 8 December 1924; Meyer, 161–62. Eckener stated that as a business proposition, Hedin's project was the more expensive of the two; Hedin would actually have to pay more to the American Zeppelin-Goodyear corporation than if he commissioned an airship directly with Luftschiffbau Zeppelin, which could build it under Allied control. Amundsen's apparent friendliness toward Eckener is somewhat surprising in view of his reportedly anti-German sentiments.

43. Brigitte Schroeder-Gudehus, "The Argument for the Self-Government and Public Support of Science in Weimar Germany," *Minerva* 10, 4 (October 1972): 553. Concerns over German scientific input into world culture were brought up in Reichstag debates.

44. PA, R32458, letter of Stockholm embassy to AA, 2 February 1925.

45. Ibid.; cf. PA, R32458, letter of AA to Christiania, Copenhagen, and Helsinki embassies, 15 December 1924; Kjaerheim, *Nansen Brev IV*, letters of Nansen to Eckener, 27 March and 5 May 1925. By January 1925 Bruns had gone on to present to the Norwegian geographical society his plans for a 1927 flight that Nansen would lead.

46. Ibid.; PA, R32458, AA circular to Scandinavian envoys, 27 January 1925. Consul Schulz-Sponholz discovered through a media source that Johann Schütte was competing for the contract. The Transportation Ministry's air section was apparently not aware of this issue.

47. PA, R32458, letter of Stockholm embassy to AA, 11 March 1925; Meyer, 160–61.

48. PA, R32804, letter of Stockholm embassy to AA, 18 February 1925.

49. PA, R32458, position paper of the German section of Aeroarctic, 20 January 1925.

50. Forman, 163; cf. Schroeder-Gudehus, 551–52.

51. PA, R32458, letter of Stockholm embassy to AA, 11 March 1925.

52. PA, R32458, letter of Moscow embassy to AA, 2 March 1925; Kjaerheim, *Nansen Brev IV*, letter of Nansen to Tschitcherin, 20 September 1924; Meyer, 161–62.

53. PA, R32458, letters of RMI to AA, 18 March 1925, and Aeroarctic to AA, 20 March 1925. The RVM laid down three main conditions for funding: (1) the airship was to be built in Germany; (2) Zeppelin could not be excluded from the construction; and (3) the Bruns plans had to remain confined to polar exploration and could not be coupled with state or air-transport goals. The RMI's conditions were similar, although they specified cooperation between Bruns and Eckener and went on to require that the expedition be led by an international scientific team with Nansen at its head.

54. PA, R32909; Dorothea Haaland, *Der Luftschiffbau Schütte-Lanz, Mannheim-Rheinau (1909–1925): Die Geschichte einer innovativen Idee als zeitlich räumlicher Prozess* (Mannheim: Institut für Landeskunde und Regionalforschung, 1987), 167–68, 174–77.

55. Eckener, 96–97.

56. PA, R32458, letter of RVM to Eckener, 15 May 1925; *Berliner Tageblatt*, 2 and 10 April 1925; *Der Tag*, 8 April 1925.

57. PA, R32804, letters of Stockholm embassy to AA, 24 April and 20 May 1925; *Stockholm Tidningen*, 29 June 1925; *Svenska Dagbladet*, 8 September 1925; Kjaer-

heim, *Nansen Brev IV,* letter of Nansen to Kohlschütter, 17 April 1925. Nansen expressed doubts concerning Bruns's willingness to compromise with the Hedin-Eckener plan. Hedin, suspicious of double-dealings, expressed puzzlement at Nansen's support of Bruns and suggested that Nansen simply wanted the airship to be built abroad so that no political complications would result from the construction decision.

58. Nansen was at the time High Commissioner for Refugees and the official representative of Norway to the League of Nations; he knew the French delegates, Pierre Flandin and Paul Boncourt.

59. PA, R32804, AA memorandum to European embassies regarding Nansen talks, 30 May 1925.

60. BA, R43II/698, 242–43; letter of Kohlschütter to RVM, 24 June 1925; Kjaerheim, *Nansen Brev IV,* letter of Nansen to Eckener, 5 August 1925.

61. *Luftfahrt,* 11 June 1925.

62. PA, R32458, letter of Eckener to AA, 10 February 1925; memorandum on Eckener-Aeroarctic meeting, 26 June 1925; Eckener had already tried to discredit Bruns on the basis of "moral right and other qualifications." The "other qualifications" likely involved Bruns's war record of some thirty flights. While honorable, this record paled in comparison with the achievements of other airship officers.

63. PA, R32458, letter of Hergesell to AA, 3 July 1925; BA, R43II/698, 271–72, letter of Eckener to Kohlschütter, 23 July 1925; *Vossische Zeitung,* 27 June 1925.

64. *Deutscher Wille,* 15 June 1925, 225–26.

65. PA, R32458, letter of Oslo embassy to AA, 6 and 8 July 1925; *Vossische Zeitung,* 16 July 1925; cf. Eckener interview, *Frankfurter Zeitung,* 26 July 1925.

66. Eckener, 97–98.

67. *Deutsche Allgemeine Zeitung,* 29 July 1925; *8 Uhr Abendblatt,* 28 July 1925.

68. PA, R32458, letters of Bruns to AA, 18 July 1925; Aeroarctic to RVM, 3 August 1925; BA, R43II/698, 268–70, letters of Kohlschütter to Federal Chancery and RVM, 11 August 1925. Aeroarctic felt that Eckener's move might be interpreted as a stab in the back to Nansen in his efforts to obtain construction clearance.

69. PA, R32809, transcription from *L'écho de Paris,* 19 June 1925.

70. PA, R32804, letters of AA to State Secretary von Schubert, 8 August 1925, and of Stockholm embassy to AA, 12 August 1925. Reporting on Nansen's recent trip to Russia, von Blücher claimed that Tschitscherin had suggested that Russia fund and build the airship, but Nansen had declined to decide on the offer.

71. *Flug,* July 1925, 10–15.

72. PA, R32804, AA memorandum, 8 September 1925. Both Hedin and Nansen attended the conference in Dessau, which received international coverage.

73. Transcript of Kohlschütter Berlin radio speech, 25 November 1925, reprinted in *Der Luftweg,* January 1926, 23–24; cf. Eckener interview, *Deutsche Tageszeitung,* 5 June 1926; *8 Uhr Abendblatt,* 8 July 1926.

74. BA, R43II/698, 290–301, letters of Aeroarctic to Chancery, AA, RMI, and Prussian Ministry of Science, Art, and Education, 28 December 1925; PA, R32804, correspondence concerning the lack of financial support; cf. Meyer, 156–59. In December 1925 Aeroarctic had pointedly reminded the German government of the international climate, which meant that only an internationally oriented group had any chance of getting permission to build an airship.

75. BA, NL42/98. Külz, then mayor of Zittau, flew aboard a special Leipzig-Zittau "charter" in July 1913. He served in both of Chancellor Luther's cabinets, from January 1925 through May 1926.

76. BA, R43II/699, 32–33, letter of RVM to RMI, 6 March 1926.

77. PA, R32459, AA memorandum, 25 April 1928.

78. Meyer, 155; Eckener, 95–96. Stresemann was reportedly angry a newspaper comment stating that the ZR 3 flight to the United States had done more for German-American relations "than ten ambassadors."

79. PA, R32458, AA internal memorandum, 27 May 1926; PA, R32459, letter of RVM to AA 23 August 1928, appendix.

80. PA, R32459, AA memorandum, 25 April 1928; Meyer, 163.

81. Breitfuss, introduction; Der Luftweg 10 December 1926, 236–39.

82. Schroeder-Gudehus, 546.

83. Ibid., 542, 544.

84. Aeroarctic, Das Luftschiff, 16–19; cf. BA, R43II/699, 160–65, "Mitgliederliste (1926)." Of the 190 members at the time of the first assembly, 6 were French. With French public interest in airships considerably dampened by the crash of the Dixmude (a Zeppelin handed over to France as war reparation), the sudden French involvement can be attributed both to interest in a scientific venture that could garner prestige and to a desire to keep an eye on German airship progress. The roster also listed Johann Schütte and August von Parseval, both earlier competitors of Count Zeppelin. Only two former airship pilots (Bruns and Blew) had a past link to the Zeppelin company.

85. PA, R32458, AA memorandum, 25 April 1928.

86. BA, Kl Erw 372 and 373.

87. Kohlschütter, in Aeroarctic, 15.

88. Fritzsche, 174–75.

89. PA, R32459; letter of Nansen to Stresemann, 14 April 1928; cf. Meyer, 163–64.

90. Luftfahrt 11 (7 June 1928): 166–68.

91. "Abschluss der Aeroarktis-Tagung," Luftfahrt, 7 July 1928, 204–5.

92. PA, R32459, AA memorandum, 25 April 1928.

93. PA, R32459, letter of RVM to AA, 23 August 1928; the requirement that Aeroarctic pay the insurance costs was confirmed at the Geneva meeting between Stresemann and Nansen. Eckener's estimate of the insurance cost included 20 per-

cent of the airship's value (4 million marks) or 800,000 marks, 25,000 per head for a crew of thirty, 200,000 for damage costs, and 250,000 for support areas in the northern regions. Furthermore, the insurance companies had just issued conditions forbidding flights beyond the Arctic Circle.

94. PA, R32459, RVM ministerial circular to other ministries, 12 May 1928; letters of RFM to RVM, 21 May 1928; letter of Paris embassy to AA, 2 July 1929. The German government particularly discouraged Bruns's plans for assistance from the French section of Aeroarctic, headed by General Delcambre. When Bruns sought the German embassy's support, it declined to assist unless Delcambre made the first move.

95. PA, R32459, RVM memorandum, 27 November 1928.

96. PA, R32459, letter of Nansen to Stresemann, 4 September 1928; PA, R32996, *Krasnoye Snamja,* 12 June 1928. Aeroarctic changed the wording on its letterhead from "Research of the Arctic by Airship" to "Research of the Arctic by Air Vehicle." Negotiations with the Soviet branch of Aeroarctic (headquartered in Leningrad) led to the establishment of a five-year commission to study plans for a polar exploration. Furthermore, an anchormast was reportedly built in Vladivostok to receive the *Graf Zeppelin* on its world flight.

97. PA, R32459, AA memorandum, 9 September 1929; cf. PA, R32988, letter of AA to RMI and RVM, 20 March 1929; PA, R32996, letter of Aeroarctic to AA, 3 January 1929. Prior to his trip to the United States, the society had asked Berlin to help Bruns, as he was not privy to business contacts in America. Bruns also tried to get the *New York Times* involved, but the paper could not match the funds Hearst had paid to obtain exclusive rights to the story.

98. WTB, 1 April 1929; PA, R32989, letter of Washington embassy to AA, 6 September 1929.

99. PA, R32459, memorandum, 7 September 1929; Schroeder-Gudehus, 560–63.

100. PA, R32459, memoranda, 7 September and 17 October 1929; Steinar Kjaerheim, *Fridtjof Nansen Brev V (1926–1930)* (Oslo: Universitetsforlaget, 1978), letter of Nansen to Bruns, 7 November 1929. Bruns brought up money issues at every ministerial gathering he attended. This behavior was all the more exasperating to those present as their positions on funding had already been clearly stated. Eventually, Ullstein got the news contract on the condition that it not impose a blackout on reports sold to other organizations.

101. WTB, 8 November 1929; cf. *8 Uhr Abendblatt,* 9 November 1929; PA, R32459, AA memorandum, 11 November 1929; cf. PA, R32991, letter of RMI to AA, 29 January 1929. The issue of operating funds appeared to lose its importance in the face of insurance companies' continuing refusal to cover a Zeppelin flight to the Arctic. WTB publicized the difficulties encountered regarding an agreement with the Deutsche Luft Pool, an association of insurance companies or departments that underwrote all flying machines in Germany.

102. PA, R32459, letter of Aeroarctic to AA, 6 December 1929. The society blamed its financial difficulties on a series of recent aviation disasters and on the failure of the Frankfurter insurance company, among the first hit by the Wall Street crash.

103. PA, R32991, letter of RMI to AA, 29 January 1930.

104. PA, R32459, letter of Külz to Minister Severing, 8 December 1929; *Deutsche Allgemeine Zeitung*, 18 December 1929.

105. LCM, Hildebrandt papers, box 14, report on a confidential conversation between Scherl representative von Euschwege and Eckener, Berlin-Schöneberg, 10 September 1929.

106. Kjaerheim, *Nansen Brev V*, letter of Nansen to Bruns, 26 October 1929; PA, R32991, letter of RMI to AA, 29 January 1930. The sudden backing out of the insurance companies had also placed the Zeppelin company in a tense situation, since both British and American insurers were concerned about whether to agree to insure the *Graf* to 100 percent of its value, as the Zeppelin company wanted it, or at only 75 percent.

107. *Geographische Zeitschrift* 36 (1930): 235, 300.

108. PA, R32459, letter of Oslo embassy to AA, 8 July 1930. Professor Adrian Mohr attacked Eckener in an interview granted to the Norwegian paper *Tidens Tegn*.

109. Kjaerheim, *Nansen Brev V*, 188–98; Detlef Brennecke, *Fridtjof Nansen* (Hamburg: Rowohlt, 1990), 125.

110. PA, R32459, letters of Eckener to Bruns, 2 September 1930, and *Aeroarctic* to AA, 3 November 1930.

111. PA, R32459, letters of *Aeroarctic* to AA, December 1930 and January 1931. The society had depended on American funding to function and to publish its journal, but this source dried up as the economic crisis deepened.

112. *Vossische Zeitung*, 12 July 1931.

113. PA, R32992 letters from Copenhagen embassy and Reykjavik consulate to AA, July 1931; cf. Carl Bruer, *Mit dem Luftschiff "Graf Zeppelin" nach Island vom 30. Juni bis 3. Juli 1931* (Goslar: n.p., 1931). According to the German general consul Schillinger in Reykjavik, the airship's visit was considered a particular honor as the nation had just celebrated its Parliament's millennium. The visit, however, also led to misunderstandings in Denmark, which twice before had expected a Zeppelin overflight that never occurred.

114. *Vossische Zeitung*, 31 July 1931.

115. BA, R43I/1450, 493–507, ministerial meeting of 21 July 1931.

116. *Exelsior*, 8 March 1931.

117. *Vossische Zeitung*, 24 July 1931.

118. Hans von Schiller, *Zeppelinbuch* (Leipzig: Bibliographisches Institut, 1938), 163.

119. Ibid.

120. Arthur Koestler, *Arrow in the Blue* (New York: Macmillan, 1952), 328. According to Koestler, Eckener landed the ship only after the mayor of Berlin agreed to pay 10,000 marks against gate receipts at Staaken Airfield.

121. Ibid., 327.

122. Ibid., 318–19; J. Gordon Vaeth, *Graf Zeppelin* (New York: Harper, 1958), 119.

123. *Südfunk*, coverage of *Graf Zeppelin* departure, 24 July 1931, *Stimmen des 20. Jahrhunderts: Der Zeppelin in Deutschland, 1900 bis 1937*, DHM DRA II 1997, compact disc.

124. Vaeth, 117–19; Umberto Nobile, *My Five Years with Soviet Airships* (Akron, Ohio: LTA Society, 1987), 24–26.

125. Ludwig Kohl-Larsen, ed., *Die Arktisfahrt des "Graf Zeppelin"* (Berlin: Union Deutsche Verlagsgesellschaft, 1931), 90–91.

126. Ibid., 199.

127. Oscar Handlin, "Science and Technology in Popular Culture," in *Science and Culture*, ed. Gerald Holton (Boston: Houghton Mifflin, 1965), 194.

128. *Vossische Zeitung*, 31 July 1931.

129. Ibid.

130. PA, R32995, letter of Eckener to RVM, 23 December 1931.

131. The International Polar Year 1932–33 was chosen to coincide with the fiftieth anniversary of the first one. Like its earlier manifestation, it was intended to be a worldwide scientific enterprise that used earth sciences to investigate both poles.

132. PA, R32995, letters of Moscow to AA, 23 December 1932, and AA to RVM and RPM, 9 January 1933. Some 30,000 gold rubles remained unpaid to the Zeppelin company for both the 1930 Russia flight and the 1931 Arctic flight. Bruns, who had returned from his Arctic trip aboard the icebreaker *Malygin*, was sent to Moscow in December 1932 as a Zeppelin representative to retrieve the funds.

133. PA, R32995, letters of RLM to AA, 26 April 1933, and AA to Moscow embassy, 12 May 1933. The Zeppelin company received 25,000 marks as payment. In the late 1920s the Zeppelin company was willing to provide technical advice for the construction of an airship in Russia on the condition that first a machine be bought from Friedrichshafen. The Soviet government, which had also floated the idea of technical cooperation with the German Trade Ministry, abandoned the talks when the payment conditions were given (one-third on order, one-third on construction, one-third on delivery). By then, General Nobile had arrived in Moscow after his disgrace in Rome. However, rumors of Russian discontent with his services led the German Air Ministry to believe that the Soviets wanted to resume talks with Zeppelin. Nothing came of it.

134. PA, R32996, letter of Bruns to RMVP, 16 February 1934.

135. PA, R32804, memorandum of AA, 17 December 1935.

136. PA, R32804, *Berliner Lokal Anzeiger*, 16 March 1934. Eckener pointed out that the idea of an airship rescue had originated "in Russian and aviation circles."

An RLM memorandum of February 1934 considered the possibility of inflating the *Graf Zeppelin* right away, but this proved to be infeasible.

137. PA, R32996, letter of Peking embassy to AA, 18 February 1935. The article forwarded claimed that the new LZ 129 under construction would make a flight to the Arctic. Similar claims circulated about trips to the Dutch West Indies and Brazil.

138. PA, R32804, AA memorandum, 17 December 1935. Eichhorn wrote for *L'avenir du Luxembourg* under the pseudonym Lucien Darel.

139. Kohl-Larsen, 14–15.

140. PA, R32996, letters of RLM to Zeppelin company, 30 August 1934; Rio embassy to AA, 8 September 1934; cf. *Berliner Lokalanzeiger,* 28 August 1934. Other attempts at applying the German airship to research included plans for a survey of the Brazilian rain forest.

141. *Die Naturwissenschaften* 20, 16 (15 April 1932): 286.

142. DeVorkin, 323.

143. Ibid., 355–58.

144. Alon Confino, *The Nation as a Local Metaphor: Württemberg, Imperial Germany, and National Memory, 1871–1918* (Chapel Hill: University of North Carolina Press, 1997), 157.

Chapter Six. Political Zeppelinism: Manipulating Airship Culture, 1933–1939

1. Hugo Eckener, *Im Zeppelin über Länder und Meere* (Flensburg: Christian Wolff, 1949), 436; Lutz P. Koepnick, "Fascist Aesthetics Revisited," *Modernism-Modernity* 6, 1 (1999): 68.

2. Raymond Firth, *Symbols: Public and Private* (Ithaca, N.Y.: Cornell University Press, 1973), 20, 22, 72–73.

3. Arnold Künzli, "Die Funktion des Symbols in der Politik," in *Welt der Symbole: Interdisziplinäre Aspekte des Symbolverständnisses,* 2d ed., ed. Gaetano Benedetti and Udo Rauchfleisch (Göttingen: Vandenhoeck & Ruprecht, 1989), 234–35.

4. Ibid., 238–39, 246; cf. Elisabeth Fehrenbach, "Über die Bedeutung der politischen Symbole im Nationalstaat," *Historische Zeitschrift* 213 (1971): 296–357; Walter Lippmann, *Public Opinion* (1922; reprint, New York: Free Press, 1965); Carlton Hayes, *Essays on Nationalism* (New York: Macmillan, 1926).

5. R. H. Hook, "Phantasy and Symbol: A Psychoanalytic Point of View," in *Fantasy and Symbol,* ed. R. H. Hook (London: Academic Press, 1979), 271.

6. Detlev Peukert, *Inside Nazi Germany: Conformity, Opposition, and Racism in Everyday Life* (New Haven: Yale University Press, 1987), 245–49.

7. Manfred Rauh, "Anti-Modernismus im nationalsozialistischen Staat," *His-*

torisches Jahrbuch 107 (1987): 94–121; cf. Hans Mommsen, "Nationalsozialismus als vorgetäuschte Modernisierung," in *Der historische Ort des Nationalsozialismus: Annäherungen,* ed. Walter H. Pehle (Frankfurt a.m.: Fischer, 1990), 31–46.

8. Rainer Zittelmann and Michael Prinz, eds., *Nationalsozialismus und Modernisierung* (Darmstadt: Wissenschaftliche Buchgesellschaft, 1991); cf. Peter Reichel, *Der schöne Schein des Dritten Reiches: Faszination und Gewalt des Faschismus* (Munich: Carl Hanser, 1991).

9. Eckener, *Zeppelin über Länder,* 432–33.

10. Peter Hayes, *Industry and Ideology: I.G. Farben in the Nazi Era* (New York: Cambridge University Press, 1987), 73, 77.

11. Rolf Italiaander, *Hugo Eckener: Ein moderner Kolumbus* (Konstanz: Stadler, 1979), 119–20. Eckener did not attend the meeting. He was traveling to Batavia, capital of the Dutch East Indies, to negotiate for the establishment of a transcontinental airship link.

12. BA, R2/5607, report from the *Deutsche Revisions- und Treuhand-Aktiengesellschaft,* 13 May 1932.

13. Claude Dornier, *Aus meiner Ingenieurlaufbahn* (Zug, Switzerland: n.p., 1966); BA, R2/5607, report from Zeppelin company to Labor Ministry, 25 January 1933.

14. The Air Commissioner's Office became the Air Ministry (Reichsluftfahrtministerium, or RLM) in the spring of 1933 and was restructured. The airship department was headed by retired airship captain Joachim Breithaupt, a World War I veteran who actually wielded little power.

15. Eckener, *Zeppelin über Länder,* 432.

16. BA, R2/5607, letter of air commissioner to work minister, 9 March 1933; Johannes Poeschel and Walter Georgii, *Ins Reich der Lüfte,* 4th ed. (Leipzig: Voigtländer, 1936), 249.

17. BA, R2/5607, letter of RLM to RFM, 21 October 1933; cf. Eckener, *Zeppelin über Länder,* 434. Following a series of complicated financial negotiations, an RLM loan of 3 million marks, taken from the 37 million marks earmarked for civilian air-defense measures, was approved. In his memoirs, Eckener claimed that this was a 2–million-mark open-ended loan/subsidy.

18. *Die Tagebücher von Joseph Goebbels,* vol. 2, ed. Elke Fröhlich (Munich: Saur, 1987), 2 July 1933 entry.

19. Peter Fritzsche, *A Nation of Fliers: German Aviation and the Popular Imagination* (Cambridge: Harvard University Press, 1992), 189; cf. Henry Cord Meyer, *Airshipmen, Businessmen, and Politics, 1890–1940* (Washington, D.C.: Smithsonian, 1991), chaps. 9 and 10.

20. *Völkischer Beobachter,* 17 October 1928; *Illustrierter Beobachter* 20 (1 October 1928): 251; 14 (6 April 1929): cover; 38 (21 September 1929): 467.

21. Cheryl Ganz, "The 'Graf Zeppelin' and the Swastika: Conflicting Symbols at

the 1933 Chicago World's Fair," *1998 National Aerospace Conference Proceedings* (Dayton, Ohio: Wright State University, 1999). The flag order approved by a puppet Reichstag applied to all flying machines and called initially for the application of the swastika on one side and imperial colors on the other. By late 1934, however, the order was changed to require the swastika on both sides.

22. Jacques Borgé and Nicolas Viasnoff, *Le Zeppelin* (Paris: Balland, 1976), 102. Evidence of what happened at this meeting, which took place in June 1933, is scarce, but several contemporaries of Eckener, including the French aeronaut and historian Charles Dollfus, claimed that Diels produced a file containing enough evidence to brand Eckener as an enemy of the state.

23. BA, NL89/8 (Brecht), letter of Eckener to Brecht, 30 November 1934. Brecht (1884–1977) had served as ministerial director in the Finance Ministry from 1927 to 1933, before his removal and exile to New York.

24. BA, R43F/4015, letter of Baden Reichsstatthalter Wagner to Chancellor's Office, 18 July 1933, and response, 2 August 1933; report on discussion between Eckener and Freiburg mayor Dr. Kerber. Kerber's own report (5 August 1933) was forwarded to Berchtesgaden at Hitler's request and passed on to Hermann Göring.

25. Leo Strauss, *De la tyrannie* (Paris: Gallimard, 1954), 42.

26. Dominique Pélassy, *Le signe nazi: L'univers symbolique d'une dictature* (Paris: Fayard, 1983), 9; cf. James D. Shand, "The Reichsautobahn: Symbol for the Third Reich," *Journal of Contemporary History* 19, 2 (April 1984): 189–200.

27. PA, R32995, letter of Rome embassy to AA, 3 April, 3 and 12 May 1933; *Schlesische Zeitung*, 30 May 1933. Italian air minister Balbo was invited on board and flew over Rome with Goebbels and German ambassador von Hassel. Goebbels's idea of using the *Graf Zeppelin* for this purpose was far from original. French and British politicians had sought to use airships for foreign policy visits at least a decade earlier.

28. PA, R32996 letter of German consulate in Geneva to AA, 24 July 1933; BA, R43II/697b, letter of Commerce Minister Oswald Lehnich to Chancery Office, 10 October 1933.

29. Peter Fritzsche, "Machine Dreams: Airmindedness and the Reinvention of Germany," *American Historical Review* 98, 3 (June 1993): 685–709.

30. Eckener, *Zeppelin über Länder*, 433. Following the initial credit advance, Goebbels offered to channel another 2 million marks to Eckener, an offer that upset Air Ministry officials.

31. Ernst Heinkel, "Das Werk der Technik," in *Deutscher Geist: Kulturdokumente der Gegenwart*, ed. Carl Lange and E. A. Dreyer (Leipzig: Voigtländer, 1933), 138.

32. *Schlesische Zeitung*, 2 May 1933.

33. Wieland Elfferding, "Von der proletarischen Masse zum Kriegsvolk," in *Inszenierung der Macht*, ed. Klaus Behnken et al. (Berlin: Dirk Nischen for NGBK,

1987), 17–50; cf. George Mosse, *The Nationalization of the Masses: Political Symbolism and Mass Movements in Germany from the Napoleonic Wars through the Third Reich* (1975; reprint, Ithaca, N.Y.: Cornell University Press, 1991), 7–10.

34. *Schlesische Zeitung*, 6 June 1933. The *Graf Zeppelin* airship photo album, published in an edition of two hundred thousand, was one of several with a nationalist theme. Later editions included Marshall Hindenburg, the *Hindenburg* airship, and of course the Führer. In each case, the sales slogan was "every German should own it even if he is not a picture collector."

35. The field had been named in honor of the count long before the Nazi party started using it for its rallies.

36. *Schlesische Zeitung*, 3 September 1933.

37. Albert Sammt, *Mein Leben für den Zeppelin*, 2d ed. (Wahlwies: Pestalozzi Kinderdorf, 1989), 56–57.

38. *Völkischer Beobachter*, 5 September 1933.

39. Walter Dornberger, *V 2: Der Schuss ins Weltall* (Esslingen: Bechtle, 1952), 75–76; cf. Werner von Braun and Bernd Ruland, *Mein Leben für die Raumfahrt* (Offenburg: Burda, 1969), 108. Hitler was visiting a rocket test-site at Kummelsdorf-West in March 1939 when he made this remark. Both Walter Dornberger and Werner von Braun noted that the Führer also minimized Count Zeppelin's legacy as inventor.

40. Eckener, *Zeppelin über Länder*, 436. While on a Zeppelin flight over a section of the Autobahn, Work Minister Todt reportedly noted that "the Führer wanted nothing to do with the airship."

41. Albert Speer, *Technik und Macht*, ed. Adalbert Reif (Munich: Bechtle, 1981).

42. BA, R43II/697b, unattributed, undated clipping [ca. 1934]; Italiaander, 120; cf. Eckener, *Zeppelin über Länder*, 509. Eckener later stated that Hitler's entourage was the source of the suggestion that the dirigible be named after him.

43. BA, R43I/585, letter of Chancellor Brüning to Eckener, February 1932; cf. Bella Fromm, *Blood and Banquets* (1943; reprint, New York: Touchstone, 1990), 219. As an aside to his request to use the airship as a political campaign tool, Brüning suggested to Eckener that LZ 129 be named after President Hindenburg.

44. Herbert D. Andrews, "Hitler, Bismarck, and History," *German Studies Review*, 14, 3 (October 1991): 510.

45. BA, NS6/216, 219, 224, 226, 229, and 231.

46. PA, R32996, RLM-AA-RMVP correspondence, June–July 1934. Such disputes often occurred where foreign journalists were concerned. An example is the case of Kazimierz Wierzynski, a Polish writer who had won a prize in the eleventh "Olympiade Lorbeer" poetry competition. The Propaganda Ministry justified inviting him on board a transatlantic flight by arguing that no Pole had ever flown on a German airship before, and the repercussions would thus be positive. The dispute went on for over a month. The transatlantic invitation was eventually bargained

down to a seat on board a "Swiss trip" flight, but then the endeavor had to be canceled because of Wierzynski's sudden illness.

47. Hans G. Knäusel, *Sackgasse am Himmel* (Bonn: Kirschbaum, 1988), chap. 3.

48. Gottfried Korff, "History of Symbols as Social History? Ten Preliminary Notes on the Image and Sign Systems of Social Movements in Germany," *International Review of Social History* 38 (1993), suppl.: 110.

49. UTD, Clara Adams papers, box 5, AC#1/77-B3F7, *DZR Betriebsmitteilungen*, September 1936, 4, Hermann Göring telegram.

50. BA, Film Archive, Richard Quaas and Hermann Stöss, *Für Ehre, Freiheit, Frieden—Zeppeline im Wahlkampf* (Syndikatfilm, 1936); *Völkischer Beobachter*, 26 March 1936.

51. *Reichssender Hamburg Funkgespräch*, 27 March 1936, *Stimmen des 20. Jahrhunderts: Der Zeppelin in Deutschland, 1900 bis 1937*, DHM DRA II 1997, compact disc.

52. BA, R78/1195, circular notices, 17 and 24 March 1936.

53. Eckener, *Zeppelin über Länder*, 486–87; BA, R78/1195, circular cable to all radio stations, 26 March 1936; Harold Dick and Douglas Robinson, *The Golden Age of the Great Passenger Airships* (Washington, D.C.: Smithsonian, 1985), 17–19, 108–9. In his haste to impress Propaganda Minister Goebbels, Captain Lehmann ordered the *Hindenburg* upward in strong wind, suddenly crushing part of the tail fin. Harold Dick was the only witness present to record the incident with his camera. The Reich radio sent out an urgent cable to all relay stations forbidding any mention of the cause of the delay; Lehmann blamed the ground crew for the incident (his version of the events is missing from the first edition of his memoirs, published in 1936, but appears in the 1937 printing).

54. BA, R78/1195, circular notice, 25 March 1936.

55. BA, ZSg 116/152, DNB report, 29 March 1936; Sammt, 137–38.

56. Dick and Robinson, 110.

57. Photocopy of propaganda leaflet, "Zeppeline im Wahlkampf," courtesy of Mr. Winfried Fischer, Bonn.

58. Eckener, *Zeppelin über Länder*, 480–83.

59. William L. Shirer, *Berlin Diary* (1941; reprint, New York: Book of the Month Club, 1987), 60.

60. Eckener, *Zeppelin über Länder*, 486–87; Dick and Robinson, 110.

61. Eckener, *Zeppelin über Länder*, 487–88.

62. BA, ZSg 117/113, clippings collection on Eckener; Eckener, *Zeppelin über Länder*, 488–89; Fromm, 219; W. E. Dodd Jr. and Martha Dodd, eds., *Ambassador Dodd's Diary, 1933–1938* (New York: Harcourt, Brace, 1941), 326, 333; cf. Meyer, 204.

63. PA, R32656, letter of Cape Town embassy to AA, 22 April 1936; cf. letter of Washington embassy to AA, 2 April 1936; William A. M. Burden, *Peggy and I* (New York: n.p., 1982), 187.

64. Jörg-M. Hormann, *Elite in the Third Reich: The History of the German Academy for Aviation Research* (Garbsen: Info-Verlag, 1988); Wolfgang Glaeser, "Die Interessenvertretung der Arbeiter im Zeppelin-Konzern von den Anfängen bis zum zweiten Weltkrieg," in *Zirkel, Zangen, und Cellon: Arbeiten am Luftschiff*, ed. Wolfgang Meighörner (Friedrichshafen: Zeppelin Museum, 1999), 86–95.

65. UTD, Rosendahl papers, box 107, folder 17, letter of Harold G. Dick to Rosendahl, April 5, 1976.

66. NARA, Berlin Documentation Center microfilm, NSDAP *Ortskartei* and *Zentralkartei*. The other party member was Max Pruss. Captains and other high-ranking Zeppelin officers who may have joined the party after the *Hindenburg* disaster are not included in this survey.

67. Ernst Willer, *Chronik der schweizer Militäraviatik* (Frauenfeld: Huber, 1990): 78; François Gross, *L'aviation populaire* (Paris: NEWZ, n.d.).

68. Emmanuel Chadeau, *Le rêve et la puissance: L'avion et son siècle* (Paris: Fayard, 1996), 163–83, 203–8; Kendall E. Bailes, *Technology and Society under Stalin and Lenin* (Princeton: Princeton University Press, 1978), 381–406.

69. Fritzsche, *Nation*, 103–31, 199.

70. Dick and Robinson, 144. The Great War ace Ernst Udet did later proceed with hookup experiments using a propelled aeroplane, but they attracted limited attention.

71. PA, R32996, *B.Z. am Mittag*, 21 April 1934.

72. Stefan Martens, "Post und Propaganda," in *Deutsche Postgeschichte: Essays und Bilder*, ed. Wolfgang Lotz (Berlin: Nicolai, 1989), 324–25.

73. Engelbert Treese, *Luftfahrtunterricht in der Volksschule* (n.p.: E. J. C. Bolckmann, 1939), 48; *Erich Beier-Lindhardt, Lese- und Arbeitsbogen für die deutsche Schule 41 a/b: Unsere Zeppelin Luftschiffe*, 3d ed. (Breslau: Heinrich Handel, 1936).

74. Christa Kamenestky, *Children's Literature in Hitler's Germany* (Athens: Ohio University Press, 1984), 151–52.

75. Karl-Heinz Ludwig, "Das nationalsozialistische Geschichtsbild und die Technikgeschichte, 1933–1945," *Technikgeschichte* 50, 4 (1983): 359–75.

76. Jeffrey Herf, "The Engineer as Ideologue: Reactionary Modernists in Weimar and Nazi Germany," *Journal of Contemporary History* 19, 4 (1984): 631–48.

77. Alfred Rosenberg, "Weltanschauung und Technik," *Deutsche Technik*, January 1938, 1–3.

78. Leonhard Adelt, *Zeppelin: Der Mann und die Idee* (Berlin: Metten, 1938), 6.

79. Wolfgang Loeff, *Drei deutsche Soldaten: Zeppelin, Schlieffen, Tirpitz* (Leipzig: Gotten, 1943), introduction; C. Lück, *Männer, Kämpfer, Sieger: Fünf Männer und Ihr Weg* (Reuthingen: Ensslin & Laiblin, 1940), 377; cf. G. Biedenkapp and H. Ult, *Unser Graf Zeppelin und sein Werk* (Braunschweig, Berlin: Westermann, 1933), introduction.

80. Loeff; cf. Heinz Luedecke, *Schiffe erobern die Luft: Erlebniss und Ergebnisse einer Arbeitgemeinschaft* (Berlin: Cecilie Dressler, 1938, 1943), 165.

81. Georg Hacker, *Die Männer von Manzell: Erinerungen des ersten Zeppelin-Kapitäns* (Frankfurt a.M.: Societätsverlag, 1936), 5; *Kyffhäuser*, 20 October 1935; cf. Luedecke, 204: "Zeppelin-type airships built abroad never achieved the perfection of the German construction," Uncle Otto tells his nephews Peter and Walter, on their way to visit the airship station in Frankfurt.

82. LCM, Hildebrandt papers, box 17, unattributed clipping, 17 February 1935.

83. Luedecke, 268–69.

84. Jay W. Baird, *To Die for Germany: Heroes in the Nazi Pantheon* (Bloomington: Indiana University Press, 1990), introduction.

85. Werner Paulman, *Ahnentafel des Grafen Ferdinand von Zeppelin* (Leipzig: Zentralstelle für Deutsche Personen- und Familien Geschichte, e.V., 1931), 3–7.

86. Tobias Engelsing, ed., *"Geliebter Ferdi, schreibe mir sobald Du kannst!"* (Konstanz: Faude, 1988); cf. Hans Rosenkranz, *Ferdinand Graf Zeppelin* (Berlin: Ullstein, 1931), 70–78.

87. Markus Werder, *Ferdinand von Zeppelin und sein Werk* (Berlin, Leipzig: Julius Beltz, 1933), 5; cf. Helmut Kayser, *Se. Exzellenz Graf Zeppelin* (Hamburg: Mordicke, 1936), introduction.

88. LCM, Hildebrandt papers, box 15, unattributed clippings on the centennial celebration at the Deutsche Akademie der Luftfahrt, 18 June 1938.

89. BA, ZSg 103/8682.

90. SWA, F25/40; cf. *Kyffhäuser*, 7 March 1937; *Berliner Börsen-Zeitung* 9 March 1937 (morning ed.); *Preussische Zeitung*, 8 March 1937. Comments on the twentieth anniversary of the count's death stand in stark contrast to those on the decennial celebration, which had emphasized the count's technological success but not his personality.

91. Hugo Eckener, *Graf Zeppelin: Sein Leben nach eigenen Aufzeichnungen und persönlichen Errinerungen* (Stuttgart: Cotta, 1938).

92. BDM, *Archivverwaltungsakten*, 0883, "Museumsgründungen 1907–1945."

93. *Frankfurter Zeitung*, 29 December 1934; cf. PA, R32996, *Berliner Lokal Anzeiger*, 28 December 1934; *Deutsche Allgemeine Zeitung*, 3 January 1935 (evening ed.).

94. PA, R32996, letter of RLM to AA, 25 July 1933; correspondence of the author with Henry Cord Meyer, 13 June 1995.

95. BA, R43II/697b, "Aussenpolitische Nachrichtensammelstelle Sonderbericht," May 1934; *Buoyant Flight* 21, 1 (November–December 1973): 4; correspondence of the author with Henry Cord Meyer, 13 June 1995. The document is something of a historiographic puzzle. Henry Cord Meyer noticed it in German archives and presented his conclusions about it at a 1973 lecture. He noted the peculiarity of the document, which, although reportedly existing in only five copies, was not designated confidential, nor was it signed. Rolf Italiaander also mentions these oddities in his 1981 biography of Eckener, adding that German archivists have

been unable to clarify at which level of the party or the government the issuing office may have functioned. References in the document to previous Special Reports do not lead anywhere, either. The document may have been issued by the SS as it sought to build a control infrastructure within the state.

96. Eckener, *Zeppelin über Länder*, 433.

97. Peter W. Brooks, *Zeppelin: Rigid Airships, 1893–1940* (Washington, D.C.: Smithsonian, 1992), 184; cf. Eckener, *Zeppelin über Länder*, 434. Depending on the source, numbers vary between 8 million and 9 million marks, which is also the final estimate of what the new airship LZ 129 actually cost.

98. DLH, Göring speech on the founding of DZR, 22 March 1935.

99. Eckener, *Zeppelin über Länder*, 504. The executive board consisted of Eckener as chairman and Lufthansa executive Carl von Gablenz and Air Ministry director Mühlig-Hoffmann as executive officers.

100. DLH, Göring speech, 22 March 1935; cf. Eckener, *Zeppelin über Länder*, 503–4. Captain Lehmann's service in the war as an airshipman was lauded, while Executive Officer Christiansen's success in breaking the blockade of German East Africa was described as a demonstration of "German virile courage."

101. DLH, Göring speech, 22 March 1935.

102. BA, ZSg 116/152, DNB report 1412, 18 September 1935.

103. DLH, DZR press release, 9 December 1935.

104. *Germania*, 30 September 1936; cf. PRO, Avia 2/1980. The comment was made by Dr. Heinz Orlovius, an RLM official who favored the joint use of the airplane and the airship.

105. Robin Higham, *The British Rigid Airship, 1908–1931: A Study in Weapons Policy* (London: G. T. Foulis, 1961), 295–313; Neville Shute, *Slide Rule* (New York: William Morrow, 1954); Laurent Wattebled, *La catastrophe du "R-101"* (Beauvais: Houdeville, 1990); Meyer, chap. 8.

106. Dick and Robinson, 85, 98.

107. ETH, Paul Jaray papers, Hs 1144, 38–51. Patent 391494, "long-range passenger airship" (filed in 1919), became the property of the Zeppelin company under contract agreement.

108. SWA, F69, "Ebel Photoalbum."

109. UTD, Clara Adams papers, box 5, AC#1/77-B3F7, manuscript, "The First *Hindenburg* Flight"; J. Gordon Vaeth, "Zeppelin Decor: The *Graf Zeppelin* and *Hindenburg*," *Journal of Decorative and Propaganda Arts* 15 (Winter–Spring 1990): 48–58.

110. *Die Woche*, 13 January 1934; Fritzsche, 187.

111. Peukert, 247.

112. *Gleisdorfer Zeitung* (Austria), 9 April 1932.

113. *Die Woche*, 13 January 1934.

114. Eckener, *Zeppelin über Länder*, 505–6.

115. Hans von Schiller, *Zeppelin: Aufbruch ins 20. Jahrhundert*, 2d ed., ed. Hans G. Knäusel (Bonn: Kirschbaum, 1988), 143; cf. Knäusel, *Sackgasse*, chaps. 2, 3. The airship's reputed comforts turned out to be rather sparse and subject to such drawbacks as gas fumes and insufficient water reserves; yet it remained preferable to the vibrating cabins of propeller airplanes.

116. BA, ZSg 116/149, DNB interview of Eckener, 6–7 March 1936.

117. P A, R32996, *La volonté indochinoise*, 29 May 1936; the newspaper reported on an American air traveler who hoped to circumnavigate the globe by airship and aeroplane.

118. Paul Schulte, *Das Wagnis des Fliegenden Paters* (Paderborn: Bonifacius, n.d.), 11–13, 91–96; Dick and Robinson, 124. Schulte was a World War I pilot who entered the priesthood after the conflict.

119. *Die Gartenlaube*, 6 January 1937, 17–18; the event took place in November 1936.

120. BA, ZSg 116/152, DNB report 31 March 1936.

121. Deutsche Zeppelin Reederei, *Airship Voyages Made Easy* (n.p., n.d.)

122. Ibid.

123. Siegfied Kracauer, "Travel and Dance," in Siegfried Kracauer, *The Mass Ornament*, ed. Thomas Y. Levin (Cambridge: Harvard University Press, 1995), 73.

124. *Göttinger Nachrichten*, 24 March 1937.

125. *German-American Commerce Bulletin* (New York) 9, 2 (May 1936). The ad was also meant to promote the Leipzig fall fair.

126. PA, R32997, minutes of RLM meeting, 20 November 1936.

127. Konrad Jarausch, *The Unfree Professions: German Lawyers, Teachers, and Engineers, 1900–1950* (New York: Oxford University Press, 1990), 157.

128. Quoted in ibid., 4.

129. SWA, F25/40; BA, R57 neu/1803.

130. SFA, E 2001 (C) Akz. 4, no. 137.

131. Ibid.

132. SFA, E 27, archive no. 23330. One officer's report suggested that in case of recurrence, he could not guarantee that his men would not fire in anger at the dirigible.

133. SFA, E 2001 (D) 1/21e, letter of Swiss president Motta to Transport and Postal Minister Pilet-Golaz, 1 May 1937. There already existed previous frictions between Switzerland and Germany because of the Swiss outlawing of the distribution of the pro-Nazi *Berliner Börsenzeitung* paper.

134. SFA, E 2001 (D) 1/21e, memorandum, 29 April 1937. The brief was apparently overlooked in light of the destruction of the *Hindenburg* a week later.

135. PA, Gesandtschaft Bern, 2473, 2477, 2478 (XIX 10a, XIX 13b, Bd.1–2). Switzerland was still recovering from the world economic crisis, which had forced a devaluation of the Swiss franc in 1936.

136. BA, ZSg 116/173, DNB, 7 May 1937; correspondence of the author with Henry Cord Meyer, 13 June 1995.

137. UTD, Rosendahl papers, box 107, folder 17, letters of Harold G. Dick to Rosendahl, 22 March and 5 April 1976.

138. *Frankfurter Zeitung*, 8 May 1937.

139. *International Military Tribunal, NCA, Julius Streicher*, vol. 2, 1675–85, quoted in Frederick M. Schweitzer, "Julius Streicher and Nazi Medievalism," *Proceedings of the Fourth Biennial Conference on Christianity and the Holocaust* (Lawrenceville, N.J.: Rider University, 1996), 153.

140. *Völkischer Beobachter*, 8 May 1937; *Das Schwarze Korps*, 27 May 1937.

141. BA, R2/5664, DZR official eulogy, May 1937.

142. PA, Gesandtschaft Bern, 2478, contains a list of funds sent by Swiss citizens, mostly from the German-speaking cantons. Although all the enclosed messages expressed sympathy for Germany, it is unclear whether they were motivated by sorrow over the *Hindenburg*'s demise or by admiration for the Nazi regime. For a comparison of public reactions in 1908 and 1937, see Guillaume de Syon, "Bangs and Whimpers: The German Public and Two Zeppelin Disasters, 1908–1937," in *Extraordinary Reactions to Extraordinary Events*, ed. Ray Browne (Bowling Green, Ohio: Popular Press, 2001), 157–65.

143. BA, R43II/697b, RMI circular letter, 13 May 1937. The *Eintopf* (pea soup with pieces of meat) campaigns were meant to incite people to frugality in order to aid the economy.

144. BA, ZSg 116/173, DNB, 8 May 1937.

145. BA, ZSg 116/173, DNB, 7 May 1937.

146. Alfred Weber, ed., *Die Letzten Drei Jahren der DZR . . .* , vol. 1 (Karlsruhe: By the editor, 1965), 54; BA, R2/5664, DZR business report, December 1943. The funds the Air Ministry received were eventually used during the war to support the families of Zeppelin workers killed in combat. Widows received 500 marks each, and children 250.

147. DLH, confidential British evaluation report on the airship, 1937; cf. PRO, Avia 1/1931, secret memorandum, 4 July 1935; Dick and Robinson, 164–65. The latter report on transatlantic airship service stated that airplanes could yield a lower ton-per-mile cost, but these had not yet been built. The airship was also considered more comfortable and could thus "drive a flying-boat service off the market." The problem of the airship was that its use required cooperation with Germany.

148. Burden, 188; Manfred Bauer and John Duggan, *LZ 130 "Graf Zeppelin II" and the End of Commercial Airship Travel* (Friedrichshafen: Zeppelin Museum, 1996), 81–97.

149. PA, R101474, letters of Rio de Janeiro embassy to AA, 13 May and 2 June 1937; letter of Buenos Aires embassy to AA, 8 September 1937; Havas news wire, 1 April 1938.

150. Bauer and Duggan, 97–99.

151. PA, R101474, letter of RLM to AA, 30 March 1938.

152. PA, R101474, secret memorandum of RLM to AA, 21 October 1937.

153. Eckener, *Zeppelin über Länder*, 543–44.

154. DLH, DZR confidential report, 3 June 1938; Eckener, *Zeppelin über Länder*, 548, 552; Bauer and Duggan, chap. 4.

155. PA, R101474, letter of Washington embassy to AA, 21 May 1938; letter of AA to RLM 24 May 1938.

156. BA, NS14/78, Reich work leader's report on September 1938 polling of 259,000 workers, 31 January 1939. The question as to who invented the rigid airship was answered correctly in 78 percent of the cases. The question on the swastika had a correct-response rate of 78.4 percent; that on the year of the Nazi seizure of power, 78.7 percent. The question on Siegfried obtained 68 percent, that on the Olympics, 77 percent, and that on Austria, 60 percent. Virtually all respondents (98.8 percent) recognized Adolf Hitler as the Führer, but many misspelled his name. The overall average rate of correct responses was 72 percent.

157. Fritzsche, *Nation*, 190.

158. *Die Tagebücher von Joseph Goebbels*, vol. 3, ed. Elke Fröhlich (Munich: Saur, 1987), 9 and 10 May 1937 entries.

159. Alfred Weber, ed., *Die Letzten drei Jahren der DZR in Dokumente und Berichten*, vol. 2 (Karlsruhe: n.p., 1965), 33.

160. *Illustrierte Beobachter: Flugzeug macht Geschichte* (Munich: Franz Eher, 1939), 6–9.

161. *Reichspost*, 23 September 1938, 5; *Völkischer Beobachter* (Viennese ed.), 23 September 1938; *Wiener Zeitung*, 23 September 1938.

162. Bauer and Duggan, 163.

163. Ibid., 167.

164. Weber, vol. 2, 64.

165. Meyer, chap. 10; Weber, vol. 2, 127–31.

166. LCM, Hildebrandt papers, box 15, unattributed clipping, 3 June 1939.

167. Bauer and Duggan, 168–69.

168. Weber, vol. 2, letter from RLM to the Zeppelin company, 20 February 1940, and secret directive from Generalluftzeugmeister, 4 March 1940; cf. Bauer and Duggan, 173.

169. Weber, vol. 2, 102–5, letter of Eckener to Max Pruss, 23 February 1940. At the time, the Zeppelin company had already been assigned to war production.

170. NASM, A3G-309150-01, Civil Aeronautics Board press release, 18 August 1941.

171. Weber, vol. 2, 127–28.

172. Eckener, *Zeppelin über Länder*, 436. In 1935, on the occasion of the foundation of DZR, Eckener had asked Göring if he would like to take a trip on an air-

ship. Göring had firmly rejected the offer, saying he had no trust in "gasbags."

173. Weber, vol. 2, 127–31. The marshall's responses to arguments from airshipmen ranged from "That cannot be" to "So?" He also rejected the argument that helium-filled airships could be used for mine and submarine detection.

174. Ludwig, 367.

175. Hacker, 200.

176. R. N. Coudenhove-Kalergi, *Revolution durch Technik* (Vienna, Leipzig: Paneuropa, 1932), 78.

177. BA, NS14/17, Heft 1, memorandum of Reichshauptstellenleiter Dr. Kurz on "Der politische Einsatz der Technik."

178. Pélassy, 176.

179. Peukert, 249.

Conclusion

1. Arno Bammé, "Technikforschung im Spiegel des Aviatik-Projektes," in *Vom Wert der Arbeit: Zur literarischen Konstitution des Wertkomplexes 'Arbeit' in der deutschen Literatur (1770–1830),* ed. Harro Segeberg (Tübingen: Niemeyer, 1991), 329.

2. Maurice Daumas, "L'histoire des techniques: Son objet, ses limites, ses méthodes," *Revue d'histoire des sciences* 22 (1969): 5–31; Philip Scranton, "Determinism and Indeterminacy in the History of Technology," in *Does Technology Drive History?* ed. Merritt Roe Smith and Leo Marx (Cambridge: MIT Press, 1994), 143–68.

3. Wiebe E. Bijker, "Do Not Despair: There Is Life after Constructivism," *Science, Technology, and Human Values* 18 (1993): 115–16.

4. Wiebe E. Bijker, Thomas Hughes, and Trevor Pinch, eds., *The Social Construction of Technological Systems* (Cambridge: MIT Press, 1987), introduction.

5. Review of Freimut Duwe, "Demokratische und Autoritäre Technik," *Technologie und Politik* 16 (Hamburg: Rowohlt, 1980), in *Technikgeschichte* 48, 4 (1980): 324–25.

6. Susan M. Gray, *Charles A. Lindbergh and the American Dilemma* (Bowling Green, Ohio: Bowling Green State University Press, 1988), 6–7.

7. Douglas Robinson, *Giants in the Sky: A History of the Rigid Airship* (Seattle: University of Washington Press, 1973), 260.

8. BA, NL221/120 (Heuss), letter of Colsman to President Heuss, 10 October 1950.

9. *Gedächtnis-Ausstellung Hugo Eckener* (Flensburg: n.p., 1978), 15.

10. ETH, Jaray papers, Hs 1145–55, letter of Eckener to Paul Jaray, 15 June 1950.

11. Jonathan Morse, "The Language of Balloon," *Raritan* 14, 1 (Summer 1994): 34–57.

12. UTD, Rosendahl papers, AC#2/77-B69-F1, correspondence regarding resumption of airship construction.

13. Steven Biel, *Down with the Old Canoe: A Cultural History of the Titanic Disaster* (New York: W. W. Norton, 1996).

14. Ulrich Greiner, "Das Ende der Unverwundbarkeit," *Die Zeit* 15 (10 June 1998): 15–16.

15. Gérard Dupuy, "Un mythe fracassé," *Libération,* 26 July 2000; *Tageszeitung,* 27 July 2000; Michael Kläsgen, "Wilde Jagd am Himmel," *Die Zeit* 31 (27 July 2000): 1.

16. Günter Grass, *My Century* (New York: Harcourt, 1999). See the chapter on 1924.

17. *Süddeutsche Zeitung,* 7 July 1996; Guillaume de Syon, "The Zeppelin Museum in Friedrichshafen," *Technology and Culture* 40, 1 (January 1999): 114–19.

ESSAY ON SOURCES AND METHODS

This bibliographic essay does not include all the materials and methods used in the preparation of this book. Below the reader will find general information on the main primary sources consulted and suggestions on important secondary sources. The term *secondary sources* is somewhat misleading, since many contemporary accounts were used as primary evidence to illustrate aspects of popular culture. For information on these, the reader should consult the chapter notes. The books and periodicals contemporary to the era covered were consulted for the most part in the following repositories: the German Federal Archives, Koblenz; the German Air and Space Society library, Bonn; Harvard University, Widener Library, Cambridge, Mass.; the Library of Congress, Washington, D.C.; the Zeppelin Museum Archives, Friedrichshafen; the Austrian State Archives, Vienna; the Austrian National Library, Vienna; and the Technology Museum, Vienna.

Archives

There are numerous archival sources on the history of German dirigibles dispersed throughout several countries, but researchers often ignore them and rely instead on secondary sources that perpetuate myths. In the United States, the National Air and Space Museum archives contain many valuable printed sources, but they are scattered among various holdings. The Library of Congress Manuscripts Division houses the papers of Alfred Hildebrandt, an aeronautical journalist who closely followed the progress of Zeppelin airships throughout their existence. Researchers interested in the connection between German and American airships as well as the public-relations aspect of the airship experience may consult the Admiral Charles E. Rosendahl papers at the University of Texas at Dallas, the Karl Arnstein papers at the University of Akron (Ohio), and the Dugan papers at the Maryland Historical Society, Baltimore. The papers of airship historian Douglas Robinson are now housed at the University of Texas at Dallas.

In Germany, the most useful archives include the Zeppelin Museum in Friedrichshafen. The Deutsches Museum is reorganizing its archival holdings, and some extremely valuable correspondence from Hugo Eckener and Alfred Colsman is available there. There are no Count Zeppelin papers as such that are accessible. Researchers can, however, consult smaller collections scattered around Germany. (Reportedly, the Zeppelin family has chosen not to make its own holdings available for the time being.) In Freiburg, the German Military Archives are helpful with regard to airship operations between 1890 and 1918. They also house the papers of airship captain Joachim Breithaupt and those of blimp designer August von Parseval. A finding aid can be ordered, as well as selected microfilms that cover the activities of military airship units. The most useful archives for sociopolitical and cultural purposes are the German Foreign Ministry Archives and the German Federal Archives. The reader should be aware, however, that following German reunification in 1990, the German Federal Archives have started to consolidate their holdings, which means that some material previously available in Koblenz may have been moved to Berlin.

Elsewhere in Europe, the Federal Institute of Technology in Zurich holds the papers of the airship designer Paul Jaray. The Archives of the French Air Force and the Archives of the French Army, both housed at Vincennes Castle near Paris, also contain helpful material on Allied airship countermeasures in World War I. The French National Air and Space Museum at Le Bourget contains substantial material, but cataloging is not complete.

The Public Records Office in Kew also has some helpful documentation. The Science Museum in London holds the papers of Sir Barnes Wallis, designer of the British R 100, which may help in comparing design approaches.

Periodicals

Periodicals vary widely in quality, and their value to the researcher depends on the nature of the research project. For example, rumors of new flying machines were legion at the turn of the century, and while such reports are a useful index of popular enthusiasm, they mean little in the technological record. Because the means of communication were still so limited, magazines and newspapers often relied on one another when updating their columns, thus fueling the spread of distorted rumors. Since early flying machines were often assimilated to the entertainment of rich eccentrics, researchers may wish to check automobile and sports magazines, many of which covered aeronautical events up to World War I.

Some periodicals stand out, however, for the quality of their reports. These include *L'aérophile*, *Allgemeine Automobil-Zeitung* (Vienna), *L'aéronautique*, *Les ailes*, *Deutsche Lufthansa Nachrichten*, *Der Flieger*, *Der Flugkapitän*, *Deutsche Luftschiffahrt*, *Deutsche Zeitschrift für Luftschiffahrt*, *Luftweg*, *Die Luftwacht*, *Luftwissen*, *Motor*, *Österreichische Flugzeitschrift*, *Technik und Kultur*, *Technik und Wirtschaft*, *Werftzeitschrift der Zeppelin-Betriebe*, *Wiener Luftschiffer Zeitung*, and *Zeitschrift für Flugtechnik und Motorluftschiffahrt*. The most helpful airship-related historical periodicals include *Zeppelin-Nachrichten*, *Buoyant Flight*, and *Gasbag Journal*.

General Studies

The field of cultural studies remains the subject of debates centering around the orthodoxy of the Adorno-Orkheimer contribution (known as the "Frankfurt school") and its contrast with Raymond Williams's pioneering work, which established the "Birmingham school." Developments in French semiotics, such as Roland Barthes, *Mythologies* (Paris: Seuil, 1957), further complicate the matter, as do recent debates on the exact nature of popular culture, but good starting points exist. These include Richard Münch and Neil J. Smelser, eds., *Theory of Culture* (Berkeley: University of California Press, 1992), and Jeffrey C. Alexander and Steven Seidmann, eds., *Culture and Society: Contemporary Debates* (New York: Cambridge University Press, 1990). The historical relevance of this approach to history is summarized in Lynn Hunt, ed., *The New Cultural History* (Berkeley: University of California Press, 1989), and Roger Chartier, *Cultural History* (Ithaca, N.Y.: Cornell University Press, 1988). For a primer on competing approaches to the study of popular or mass culture, see John Storey, *An Introductory Guide to Cultural Theory and Popular Culture* (Athens: University of Georgia Press, 1993); Chandra Mukerji and Michael Schudson, *Rethinking Popular Culture* (Berkeley: University of California Press, 1991); and the series of debates in *Past and Present* 132 (August 1991): 131–49; 133 (November 1991): 204–13; 135 (May 1992): 189–208; and 141 (November 1993): 207–19. The *Journal of Popular Culture* also helps in evaluating the widespread dimensions of the field.

In relation to Germany proper, Russell A. Berman, *Cultural Studies of Modern Germany: History, Representation, and Nationhood* (Madison: University of Wisconsin Press, 1993), and Rob Burns, ed., *German Cultural Studies: An Introduction* (New York: Oxford University Press, 1995), help clear a path to the literature available on the subject. Other works exist that focus more specifically on one of the political eras covered in this book.

Determining the Dynamics of a Symbol

Throughout this book, it was necessary to rely upon many approaches and sources to test assumptions, confirming some while rejecting certain earlier historical findings derived from airship myths. Such myths derived less from the machine's actual technical performance than from its symbolic importance and how people perceived it.

The Zeppelin brought about a synthesis of visual experience, imagination, and ideas reflecting the circumstances and place of occurrence of the event. *Zeppelin-*

ism, a term at first (in 1910) used negatively, came to connote a multitude of positive signs and meanings. Unlike other instruments of modern technology with a direct application or function, the airship assumed its instrumental role not through its transportation or military potential but rather through its symbolic power. All these associations depended strongly on context.

Technocultural icons cannot be understood in isolation but require a careful linking to contemporary factors found in sociocultural, political, and technical history. Only in this way can one clarify certain emblematic values of the airship. Of course, the manipulation of the symbol at various stages of its evolution may not be discounted, but such manipulation is intrinsic to the evolution of any emblem. A technocultural icon such as the Zeppelin brings together a series of loosely related elements. In a wider framework, then, the symbol may also combine a neutral and a biased stance, as it absorbs different or opposing perceptions into a common meaning; in this sense, it represents what David Nye, *American Technological Sublime* (Cambridge: MIT Press, 1994), and others define as the "technological sublime" as it mediates between the different opinions surrounding a single artifact.

Andrew Feenberg, *Critical Theory of Technology* (New York: Oxford University Press, 1991), argues that individuals are themselves an integral part and a factor of technology. In the case of the airship, individuals acted at several levels. The first level involves the image of the inventor and of the promoter. Both Count Zeppelin and Hugo Eckener were essential to the instrumentalization of the airship in German culture, but as such they themselves became part of the process, achieving levels of recognition comparable to those the dirigible enjoyed. A second way in which individuals were involved concerns the role official actors played in the promotion or demotion of the symbol. The dynamic of governmental manipulation of the airship is an important part of both its success and its demise. The masses embodied the third level of individual involvement, the most pervasive and difficult to define. In the case presented here, popular opinions were expressed mostly through the mediation of the press. Although the procedure may taint the value of the record, popular action and individual recollections help balance newspaper accounts by contextualizing the reports.

Paul Rosen, *The Social Construction of Mountain Bikes: Technology and Postmodernity in the Cycle Industry, Social Studies of Science* 23 (1993), shows that discovering a machine's multiple cultural meanings requires going beyond established approaches to the social history of technology. Count Zeppelin's airships were but a transitory technology, one step toward the mastery of human flight; their stunning impact on the public, however, depended on factors beyond engineers' and businessmen's control.

As this study has shown, the phenomenon of the technological sublime is not limited to the North American context but may apply to other national emblems of technical and cultural achievement. However, as Arnold Pacey, *The Culture of Tech-*

nology (Cambridge: MIT Press, 1983), notes, not all big machines become symbols of great cultural achievement the way cathedrals in the Middle Ages did. The transformation of an object into an icon depends upon whether it is defined as religious, commercial, or technical. Marshall Fishwick and Ray B. Browne, eds., *Icons of Popular Culture* (Bowling Green, Ohio: Bowling Green University Popular Press, 1970), shows that one theme common to all definitions of an icon is that it merges images and ideas into a single point of reference.

Although the variables of *Zeitgeist* (contemporary mood) go a long way toward explaining the success or failure of a technological icon, the understanding of that process also requires an acknowledgment and appreciation of the machine's workings. The concept of a theory of technology that would embody both the "black box" (that is, the inner workings of the machine) and the ways in which technology and society influence one another remains a central theme in studies of technology and culture. Early attempts at surveying the social history of technology include Siegfried Giedion, *Mechanization Takes Command: A Contribution to Anonymous History* (New York: Oxford University Press, 1948), which, for lack of an appropriate theoretical construct, limits itself to a concept of machines as "fragments" of "anonymous history": mechanization becomes an anonymous agent—organic, structural, or functional, depending on time and circumstances—that contributes to the ongoing "dynamic equilibrium."

Whatever approach to the history of technology the reader prefers, certain basic texts provide an important foundation for the development of new (or shifting) paradigms: Thomas S. Kuhn, *The Structure of Scientific Revolutions* (Chicago: University of Chicago Press, 1962); Wiebe E. Bijker, Thomas Hughes, and Trevor Pinch, *The Social Construction of Technological Systems* (Cambridge: MIT Press, 1987); and Leo Marx, *The Machine in the Garden* (New York: Oxford University Press, 1964). George Basalla, *The Evolution of Technology* (New York: Cambridge University Press, 1988), explores the concepts of diversity, continuity, novelty, and selection. Of the general issues surrounding approaches to the history of technology, the first concerns the relevance of technology to history. See Stephen H. Cutliffe and Robert C. Post, eds., *In Context: History and the History of Technology* (Bethlehem, Pa.: Lehigh University Press, 1989), and Merritt Roe Smith and Leo Marx, eds., *Does Technology Drive History?* (Cambridge: MIT Press, 1994). On the intersection of the history of technology and business history, see Paul Uselding, "Business History and the History of Technology," *Business History Review* 54, 4 (Winter 1980): 443–58; David A. Hounshell, "Hughesian History of Technology and Chandlerian Business History: Parallels, Departures, and Critics," *History and Technology* 12 (1995): 205–24; and Hans Liudger Dienel, "Sociological and Economic Technology Research: A Guideline for the History of Technology?" *ICON* 1 (1995).

The study of the culture of technology is a rich, though often anarchical, field of research that varies according to the artifact studied. Langdon Winner, *The Whale and the Reactor* (Chicago: University of Chicago Press, 1986), provides a central foundation to the notion of the "social shaping of technology"; on the shortcomings of this approach and ways to circumvent them, see Hugh Mackay and Gareth Gillespie, "Extending the Social Shaping of Technology Approach: Ideology and Appropriation," *Social Studies of Science* 22 (1992): 685–716. For a different method of analyzing technology and culture, see the ongoing work of Wiebe Bijker and others on the notion of *The Social Construction of Technological Systems* (cited earlier in this section) and related case studies in *Shaping Technologies/Building Society: Studies in Sociotechnical Change* (Cambridge: MIT Press, 1992). Challenges to this "social constructivist approach" include Paul Rosen, *The Social Construction of Mountain Bikes* (also cited earlier). Michael Adas analyzes Western writings in *Machines as the Measure of Men: Science, Technology, and Ideologies of Western Dominance* (Ithaca, N.Y.: Cornell University Press, 1989). Mark L. Greenberg and Lance Schachterle, eds., *Literature and Technology* (Bethlehem, Pa.: Lehigh University Press, 1992), situates well the possible uses of literary texts for understanding the culture of technology.

Studying the symbolic roles of technology is a relatively new endeavor. David Nye's *American Technological Sublime* (cited earlier) and Eric Schatzberg's studies (see below) come closest to considering these notions in relation to machines. One of the most provocative questions associated with symbolism is the relationship between the symbol's intended message and its actual representation at the political and social level. These issues are discussed in Raymond Firth, *Symbols: Public and Private* (Ithaca, N.Y.: Cornell University Press, 1973); George Mosse, *The Nationalization of the Masses: Political Symbolism and Mass Movements in Germany from the Napoleonic Wars through the Third Reich* (1975; reprint, Ithaca, N.Y.: Cornell University Press, 1991); and Dominique Pélassy, *Le signe nazi: L'univers symbolique d'une dictature* (Paris: Fayard, 1983). Elisabeth Fehrenbach, "Über die Bedeutung der politischen Symbole im Nationalstaat," *Historische Zeitschrift* 213 (1971): 296–357, offers a critique of symbols as tools of nationalist propaganda and touches on a decades-old debate about the value of national symbols. The arguments in this debate range from Walter Lippmann's view that the symbol as stereotype overshadows individual will and strengthens group direction, to Carlton Hayes's defense of the symbol as a product not of elite propaganda but of the secularization of religious culture. See Walter Lippmann, *Public Opinion* (1922; reprint, New York: Free Press, 1965), and Carlton Hayes, *Essays on Nationalism* (New York: Macmillan, 1926).

Cultural studies of aeronautics have increased substantially in recent years, evolving from an initial focus on the relationship between technology and literature, as reflected in Laurence Goldstein, *The Flying Machine and Modern Literature* (Bloomington: Indiana University Press, 1986), and William C. Carter, *The Proustian Quest* (New York: New York University Press, 1992). Peter Fritzsche, *A Nation of Fliers: German Aviation and the Popular Imagination* (Cambridge: Harvard University Press, 1992), offers the best synthesis of the flying-culture phenomenon in Germany in the first third of the twentieth century. Robert Wohl, *A Passion for Wings: Aviation and the Western Imagination, 1908–1918* (New Haven: Yale University Press, 1994), the first volume of a projected trilogy, is an admirable study of the artistic perception and expression of the new technology and gives a clearer sense of the deep impact the airplane had beyond engineering, business, and political circles. Joseph Corn, *The Winged Gospel: America's Romance with Aviation, 1900–1950* (New York: Oxford, 1983), also proposes a cultural approach to the American case. Emmanuel Chadeau, *Le rêve et la puissance: L'avion et son siècle* (Paris: Fayard, 1996), attempts to synthesize a framework that places the flying machine in a worldwide context. Discussions contrasting cultural influences with engineering decisions are far fewer, though Eric Schatzberg, *Wings of Wood, Wings of Metal: Culture and Technical Choice in American Airplane Materials, 1914–1945* (Princeton: Princeton University Press, 1998), succeeds remarkably well.

Documenting the Airship

Aviation history has for the most part been the territory of enthusiasts. As James Hansen points out, though, a trend toward more context and less technical focus has emerged ("Aviation History in the Wider View," *Technology and Culture* 89, 3 [July 1989]: 643–65). Researchers should not discount studies by nuts-and-bolts specialists, however, for many of these works are treasure-troves of detail. While some rely on secondary sources and should therefore be used with caution, others are serious original investigations that help set the factual record straight.

As for airship histories, general sources are plentiful, even in the English language. Henry Cord Meyer, *Airshipmen, Businessmen, and Politics, 1890–1940* (Washington, D.C.: Smithsonian, 1991), provides the best overview of the political context surrounding airship matters in Germany as well as other Western nations. As this book goes to press, a new volume written by Meyer in cooperation with John Duggan, *The Airship in International Affairs* (New York: St. Martin's, 2001), has been announced, as has Dale Topping, *When Giants Roamed the Sky*, ed. Eric Brothers (Akron: University of Akron Press, 2001). Books by Douglas Robinson, a pioneer of airship historiography, offer an ideal starting point for those interested in airship operations. His *Giants in the Sky: A History of the Rigid Airship* (Seattle: University of Washington Press, 1973) contains a particularly good narrative of the airship era, although certain chapters are a bit outdated. Robin Higham, *The British Rigid Airship, 1908–1931: A Study in Weapons Policy* (London: G. T. Foulis, 1961), is a helpful comparative element in the study of German airships. Other general studies include Peter W. Brooks, *Zeppelin: Rigid Airships, 1893–1940* (Washington, D.C.: Smithsonian, 1992).

Among works in German, the illustrated memoirs of several airshipmen can help clarify the nature of the Zeppelin fraternity and give further information about the Zeppelin era as a whole. These include Albert Sammt, *Mein Leben für den Zeppelin*, 2d ed. (Wahlwies: Pestalozzi Kinderdorf, 1989), and Hans von Schiller, *Zeppelin: Aufbruch ins 20. Jahrhundert*, 2d ed., ed. Hans G. Knäusel (Bonn: Kirschbaum, 1988). Older works, such as Ernst Lehmann's *Auf Luftpatrouille und Weltfahrt* (Berlin: Wegweiser, 1936), also contain some helpful general information. The most encompassing reference work on airships (of all types) is Dorothea Haaland et al., eds., *Leichter als Luft-Ballone und Luftschiffe* (Bonn: Bernard & Graefe, 1997).

Imperial Germany

Biographies of Count Zeppelin abound, though no definitive account of the man's life exists. The most informative works include Hugo Eckener, *Graf Zeppelin: Sein Leben nach eigenen Aufzeichnungen und persönlichen Errinerungen* (Stuttgart: Cotta, 1938), and Wolfgang Meighörner, ed., *Der Graf, 1838–1917* (Friedrichshafen: Zeppelin Museum, 2000). Other studies either are extremely brief or intermix fact and hagiography, though they may contain some good anecdotes. See Rolf Italiaander, *Ferdinand Graf von Zeppelin* (Konstanz: Stadler, 1986); Michael Bélafi, *Graf Ferdinand von Zeppelin*, 2d ed. (Leipzig: Teubner, 1986); and Lutz Tittel, *Graf Zeppelin: His Life and His Work*, trans. Peter A. Schmidt (Friedrichshafen: Zeppelin Museum, 1995). None, however, offers a critical appraisal of the inventor and his motivations, although Henry Cord Meyer, "Militarismus und Nationalismus in Graf Zeppelins Luftschiff Idee: Eine Studie zum Thema psychologischer Kompensation," *Zeppelin Jahrbuch* (Friedrichshafen: Zeppelin Museum, 1998), represents a small step in that direction.

On early Zeppelin airships, one may consult Hans G. Knäusel, *LZ 1, der erste Zeppelin: Geschichte einer Idee, 1874–1908* (Bonn: Kirchbaum, 1985); Wolfgang Meighörner, "... *der Welt die Wundergabe der Beherrschung des Luftmeeres schenken": Die Geschichte des Luftschiffs LZ 2* (Friedrichshafen: Zeppelin Museum, 1991); and Lutz Tittel, *Die Fahrten des LZ 4, 1908* (Friedrichshafen: Zeppelin Museum, 1983). Jürgen Eichler, *Luftschiffe und Luftschiffahrt* (Berlin: Brandenburgisches Verlagshaus, 1993), helps clarify the contextual history of Zeppelins in relation to other airship solutions, domestic and international. Jeannine Zeising, "Reich und Volk für Zeppelin! Die Journalistische Vermarktung einer technologischen Entwicklung," in *Wissenschaftliches Jahrbuch 1998* (Friedrichshafen: Zeppelin Museum, 1998), 67–227, nicely frames the relationship of Zeppelin marketing to press coverage. Karl Clausberg, *Zeppelin: Die Geschichte eines unwahrscheinlichen Erfolges* (1979; reprint, Augsburg: Weltbild, 1989), offers the best mix of fact and analysis concerning the popular enthusiasm for airships.

On the early Zeppelin company, Alfred Colsman, *Luftschiff voraus!* (1933;

reprint, Munich: Zuerl, 1983), is highly informative but stops in the early Weimar era. Hans G. Knäusel, *Unternehmen Zeppelin* (Bonn: Kirschbaum, 1994), offers a brief overview of the Zeppelin concern. The situation of workers at the Zeppelin company has yet to be thoroughly investigated, but steps in that direction include Heike Vogel, *"Suche ein nettes Zimmer": Die Zeppelin-Wohlfahrt GmbH und der Wohnungsbau in Friedrichshafen* (Friedrichshafen: Zeppelin Museum, 1997), and Wolfgang Meighörner, ed., *Zirkel, Zangen, und Cellon: Arbeit am Luftschiff* (Friedrichshafen: Zeppelin Museum, 1999).

The impact of industrialization on European society and culture in the nineteenth century is the subject of hundreds of literary and historical studies. The most useful that focus on technology and its impact on communication and society are Wolfgang Schivelbusch, *The Railway Journey* (Berkeley: University of California Press, 1986), and Stephen Kern, *The Culture of Time and Space, 1880–1918* (Cambridge: Harvard University Press, 1983).

With regard to Germany proper, partisans of a historical "special path" (*Sonderweg*) argue that there is a fundamental tension between "civilization" and "culture" in German society due to the lack of a successful revolution; this fact affected industrialization and the development of democracy. The tension is explored in Werner Koene, "On the Relationship between Philosophy and Technology in the German-Speaking Countries," in *The History and Philosophy of Technology*, ed. G. Bugliarello and Dean B. Doner (Urbana: University of Illinois Press, 1979). Challenges to the notion of *Sonderweg* are best embodied in David Blackbourne and Geoff Eley, *The Peculiarities of German History* (New York: Oxford University Press, 1984), and Richard J. Evans, *Rethinking German History: Nineteenth Century Germany and the Origins of the Third Reich* (London: Allen & Unwin, 1987), and further illustrated in Geoff Eley, ed., *Society, Culture, and the State, 1870–1930* (Ann Arbor: University of Michigan Press, 1995). In terms of industrialization, consult Eric Dorn Brose, *The Politics of Technological Change in Prussia: Out of the Shadow of Antiquity, 1809–1848* (Princeton: Princeton University Press, 1993). Kurt Möser, "Zur Theorie und Praxis der Technikthematisierung bei Max Eyth," *Technikgeschichte* 52, 4 (1985): 313–27, demonstrates that establishment literature had begun to accept and incorporate technological themes by the turn of the century.

The best introduction to imperial Germany appears in Volker Berghahn, *Imperial Germany, 1871–1914: Economy, Society, Culture, and Politics* (Providence, R.I.: Berghahn, 1994). Popular nationalist associations have become the subject of several studies in recent years. The most helpful include Roger Chickering, *We Men Who Feel Most German: A Cultural Study of the Pan-German League, 1886–1914* (Boston: Allen & Unwin, 1984), and Thomas Rohkrämer, *Der Militarismus der "kleinen Leute"* (Munich: R. Oldenbourg, 1990). Michael Hughes, *Nationalism and Society: Germany, 1800–1945* (London: Edward Arnold, 1988), is a useful explanation of the variety of "nationalisms" that Germany experienced; further illustrations of this notion appear in Celia Applegate, *A Nation of Provincials: The German Idea of Heimat* (Berkeley: University of California Press, 1990), and Alon Confino, *The Nation as a Local Metaphor: Württemberg, Imperial Germany, and National Memory, 1871–1918* (Chapel Hill: University of North Carolina Press, 1997).

World War I

One of the great challenges to writing any history of World War I in the air is the fact that the official accounts often distorted the record and the postwar literature simply added new layers to the mythology. While such evidence constitutes a gold mine of cultural history, it is essential to compare it with archival material. Dominick Pisano et al., *Legend, Memory, and the Great War in the Air* (Seattle: University of Washington Press, 1992), traces the evolution of the airplane and pilot myths, while Michael Paris, *Winged Warfare: The Literature and Theory of Aerial Warfare in Britain, 1859–1917* (Manchester: Manchester University Press, 1992), explains the influence of fiction on early notions of the airship as a weapon of war. *Die Luftflotte,* which was the periodical of the German Air League, gives a clearer sense of popular expectations of the airship in war. The most helpful historical syntheses of the air war are Lee Kennett, *The First Air War, 1914–1918* (New York: Free Press, 1991), and John H. Morrow, *The Great War in the Air: Military Aviation, 1909–1921* (Washington, D.C.: Smithsonian, 1993). In addition, several specialized magazines, such as *World War I Aero* and *Over the Front,* though tending toward technical and military descriptions, include helpful historical discussions.

As Hans Knäusel once noted, a morbid glorification usually stands out in accounts of the airship's impact in war, and analysis is meager. This has begun to change. The standard work on German airships is Douglas Robinson, *The Zeppelin in Combat,* 3d ed. (Seattle: University of Washington Press, 1980). For more technological information, consult Paul Schmalenbach, *Die deutschen Marine-Luftschiffe* (Herford: Koehler, 1977), and John Provan, "The German Airships in World War I" (master's thesis, Technische Hochschule Darmstadt, 1994). Wolfgang Meighörner, *Wegbereiter des Weltluftverkehrs wider Willen: Die Geschichte des Zeppelin-Luftschifftyps "w"* (Friedrichshafen: Zeppelin Museum, 1992), offers a valuable contrast between airship engineering limitations and military requirements.

Airship raids on England have received the most historical coverage. The only published German memoir to focus on this subject exclusively is Horst von Buttlar Brandenfells, *Zeppeline gegen England* (Zurich, Leipzig: Almathea, 1931), also available in a translation: *Zeppelins over England* (New York: Harcourt, Brace, 1932). Sources on the view "from below" are far more numerous and include Barry D. Powers, *Strategy without Slide-Rule* (London: Croom Helm, 1976); Christopher Cole, *The Air Defence of Britain, 1914–1918* (London: Putnam, 1984); H. G. Castle, *Fire over England* (London: Secker & Warburg, 1982); and Neville Jones, *The Origins of Strategic Bombing* (London: Kimber, 1973). The best work, because it incorporates military as well as social history, is Raymond L. Rimell, *Zeppelin! A Battle for Air Supremacy in World War I* (London: Conway, 1984).

Studies of caricatures and related propaganda of the Great War are numerous but heavily focused thematically. The best overall collection of cartoons for the first half of the conflict remains John Grand Carteret's multivolume *Verdun: Images de guerre* (Paris: Chapelot, 1916). Good contemporary introductions to the topic include Franz W. Seidler, *Das Militär in der Karikatur: Kaiserliches Heer, Reichswehr, Wehrmacht Bundeswehr und Nationale Volksarmee im Spiegel der Pressezeichnung* (Munich: Bernard & Graefe, 1982), and Eberhard Demm and Tilman Koops, *Karikaturen aus dem ersten Weltkrieg* (Koblenz: Bundesarchiv, 1990). On the postcard's place in the Great War, see Marie-Monique Huss, *Histoires de famille: Cartes postales et culture de guerre* (Paris: Noesis, 2000).

Weimar Germany

With the exception of Peter Kleinheins, *LZ 120 "Bodensee" und LZ 121 "Nordstern": Luftschiffe im Schatten des Versailler Vertrages* (Friedrichshafen: Zeppelin Museum, 1994), which discusses the Zeppelin projects that immediately followed the Great War, most accounts of the airship in the Weimar era carry on into the Third Reich. Douglas Robinson and Charles Keller, *Up Ship! U.S. Navy Rigids, 1919–1935* (Annapolis, Md.: Naval Institute Press, 1982), is very helpful for understanding the nature of German-American negotiations on the construction and delivery of the ZR 3 airship. Possibly the best account of the transport ships, Harold G. Dick and Douglas Robinson, *The Golden Age of the Great Passenger Airships Graf Zeppelin and Hindenburg* (Washington, D.C.: Smithsonian, 1985), beautifully blends public events and their behind-the-scenes explanations, although it is at times heavily technical. Eugen Bentele, *Ein Zeppelin-Maschinist erzählt* (Friedrichshafen: Zeppelin Museum, 1990), clarifies the role of airmen in Zeppelin operations.

Information on Hugo Eckener as the architect of the airship revival, although plentiful in press accounts, is harder to come by in book form. Hugo Eckener, *Im Zeppelin über Länder und Meere* (Flensburg: Christian Wolff, 1949), constitutes a kind of memoir of airship activities during the interwar years. Highly informative at times, it does not, however, include material before 1918. (The English translation, although helpful, is abridged.) Rolf Italiaander, *Ein Deutscher namens Eckener* (Konstanz: Stadler, 1981), is the only biographical account that avoids romantic exaggeration, preferring to rely on reprinted correspondence that reflects a positive image of its subject.

The intricacies of the Aeroarctic episode are covered in part in Henry Cord

Meyer, *Airshipmen, Businessmen, and Politics* (cited earlier under "Documenting the Airship"), and in Leonid Breitfuss, *Aeroarctic* (Gotha: Justus Perthes, 1927). Further information comes to light through Fridtjof Nansen's correspondence, published in Steinar Kjaerheim, ed., *Fridtjof Nansen Brev IV (1919–1925)* (Oslo: Universitetsforlaget, 1966) and *Fridtjof Nansen Brev V (1926–1930)* (Oslo: Universitetsforlaget, 1978). Aeroarctic's journal *Arktis* also provides insight into the group's goals and workings.

The significance of Weimar modernism and its relation to the failure of German democracy are well documented and heavily disputed. Jeffrey Herf, *Reactionary Modernism: Technology, Culture, and Politics in Weimar and the Third Reich* (Berkeley: University of California Press, 1985), discusses the issue of technology and modernist culture. Detlev Peukert, *The Weimar Republic: The Crisis of Classical Modernity* (New York: Hill & Wang, 1992), offers a useful contextual counterpoint. For a review of the scholarly debate on the failure of the republic, see Peter Fritzsche, "Did Weimar Fail?" *Journal of Modern History* 68 (September 1996): 629–56.

On war literature and its interaction with public memory, see Michael Gollbach, *Die Wiederkehr des Weltkrieges in der Literatur: Zu den Frontromanen der späten zwansiger Jahre* (Kronberg T.s.: Scriptor, 1978), and Paul Fussell, *The Great War and Modern Memory* (New York: Oxford University Press, 1975). A summary of a recent debate over the selection and interpretation of such sources appears in Lynn Hanley, *Writing War: Fiction, Gender, and Memory* (Amherst: University of Massachusetts Press, 1991). On mass entertainment, the special issue of *New German Critique* 51 (Fall 1990) presents examples of Weimar popular culture.

The Weimar-era fascination with technology and speed and their implications is best presented in Michael Neufeld, "Weimar Culture and Futuristic Technology: The Rocketry and Spaceflight Fad, 1923–1933," *Technology and Culture* 31, 4 (October 1990): 725–52, and Adelheid von Saldern, "Cultural Conflicts, Popular Mass Culture, and the Question of Nazi Success: The Eilenriede Motorcycle Races, 1924–1939," *German Studies Review* 15, 2 (May 1992): 317–38. On the "rail Zeppelin" fascination, see Sigfried von Weiher, "Franz Kruckenbergs Lebenswerk, 1882–1982," *Kultur und Technik* 6, 4 (December 1982): 226–30.

Nazi Germany

The most concise social history that captures the complexities of life in the Third Reich is Detlev Peukert, *Inside Nazi Germany: Conformity, Opposition, and Racism in Everyday Life* (New Haven: Yale University Press, 1987). It includes a discussion of a central question raised in the past decade: whether the Third Reich was a modernizing force, as opposed to a reaction to modernity. On the one hand, historians like Peukert suggest that the Reich was a pathological manifestation of modernization. Thomas P. Hughes, "Ideology for Engineers," in *Ideology, Bureaucracy, and Genocide,* ed. H. Friedlander and S. Milton (Milwood, N.Y.: Kraus, 1980), emphasizes that "edited" Nazi ideology made itself attractive to engineers and technicians alike. Other historians, however, suggest that there were positive manifestations of the ideology, from highways to television. This disturbing assessment of "happy Nazism" appears in the works of Peter Reichel, *Der schöne Schein des Dritten Reiches: Faszination und Gewalt des Faschismus* (Munich: Carl Hanser, 1991), and Rainer Zittelmann and Michael Prinz, eds., *Nationalsozialismus und Modernisierung* (Darmstadt: Wissenschaftliche Buchgesellschaft, 1991). Two other works are helpful to considering the evolution and differences of design and technology between Weimar and the Third Reich: Uwe Westphal, *Werbung im Dritten Reich* (Berlin: Transit, 1989), seeks to explain the changes that occurred in advertising as it became part of propaganda efforts, while Winfried Nerdinger, *Bauhaus-Moderne im Nationalsozialismus* (Munich: Prestel, 1993), discusses the selective use of "degenerate" Bauhaus style in Nazi Germany. Jay W. Baird, *To Die for Germany: Heroes in the Nazi Pantheon* (Bloomington: Indiana University Press, 1990), and Gilmer W. Blackburn, *Education in the Third Reich: Racism and History in Textbooks* (Albany: SUNY Press, 1985), clarify the selective utilization of history in a dictatorship.

Discussions of Nazi modernity of course entail a consideration of specific totalitarian attitudes toward technology in peacetime. Multiple approaches to this subject and many helpful observations may be found in Paul R. Josephson, *Totalitarian Science and Technology* (Atlantic Highlands, N.J.: Humanities Press, 1996); Eric Dorn Brose, "Generic Fascism Revisited: Attitudes toward Technology in Germany and Italy, 1919–1945," *German Studies Review* 10, 2 (1987): 273–97; Bernard Bellon, *Mercedes in Peace and War: German Automobile Workers, 1903–1945* (New York: Columbia University Press, 1990); Monika Renneberg and Marc Walker, *Science, Technology, and National-Socialism* (New York: Cambridge University Press, 1994); and Michael J. Neufeld, *The Rocket and the Reich* (New York: Free Press, 1995).

INDEX

Eckener, Hugo 1, 140–41, 146–47, 188, 206, 210; aging, 143, 201; as airship commander, *122*, 132, 163; and arctic project, 156–57, 161–62, 165–66, 170; biographer of Count Zeppelin, 186; conflict with Nazi officials, 175– 76, 182; conflict with Weimar officials, 126; critic of reparations, 142; death, 207; disagreement with Colsman, 107, 124, 128; distrust of Bruns, 153, 156; reactions to *Hindenburg* disaster, 195–96; run for presidency, 145; in World War I, 105, 111. *See also* "Zeppelin-Eckener *Spende*"

Edison, Thomas, 108

Edward VII (King of England), 31

Eichhorn, Lucien, 169

Einem, General Karl von, *28*, 34

Einstein, Albert, 121, 149

Eliade, Mircea, 203

Ellsworth, Lincoln, 157, 167

Emergency Association of German Science, 160

Engines. *See* propulsion technology

Enlightenment, 11, 38

Ersatz Deutschland airship, *70*

Etrich, Igo, 52

Euler, August, 63

Fédération aéronautique internationale, 182

Flemming, Hans, *122*

Fleurus, battle of, 13

Fogg, Phileas, 191

Forain, J. L., 67

Ford, Ford Madox, 99

Ford, Henry, 114, 127

France, 109, 114, 122; aerial competition with, 16, 19, 30, 62, 72, 79–80

Franco-Prussian war, 12, 185; and ballooning 15, 30

Frankfurt am Main, 187

Franz-Josef, Emperor, 53, 55, 141, 149

Franz-Josef Land, 149–67

Freund, Leo, *122*

Friedrichshafen, 1, 23, 48, 58, 62, 192; factory in, 60, 114; line to Berlin, 113. *See also* Zeppelin airships; Zeppelin company

fundraising, early efforts, 16, 25–27, 32; popular, 38, *43*, 44, 196. *See also* "Zeppelin-Eckener *Spende*"

futurism, 67

futuristic novels. *See* Martin, Rudolf; Utopia; Verne, Jules; Wells, H. G.

Ganswindt, Hermann, 11

Gartenlaube, 23, 66

gasbag, 19, 21, 85, 118. *See also* ballonet

Gemmingen, Freiherr von, 115

gender and flying, 64–65, 185

German Airmen's League, 49

German Airport League, 183

German Airship League, 78, 100

German army: funding of airship, 34, 57; tensions with Count Zeppelin, 17, 29, 57, 82

German Aviation Insurance Consortium, 165

German Colonial Society, 25

German Engineers' Association. *See* VDI

German Foreign Ministry, 120–21, 123, 126, 155, 159–60

German House of Flyers, 197

Germania, 140, 180

German navy, 82, 85, 103, 106

German parliament. *See* Reichstag

Germans abroad, 123

Gianozzo (balloonist), 51

Gibbs, Philip, 97

Giffard airship, *9*

Giffard, Henry, 9–10

Gleichschaltung, 177, 191, 194

Goebbels, Joseph, 169, 175–76, 182, 187, 199. *See also* Ministry of Propaganda

Goodyear Rubber and Tire Company, 116, 207

Göring, Hermann, 174, 187, 197, 199, 202

Graf Zeppelin aircraft carrier, 186
Graf Zeppelin airship (LZ 127), 2, 112, 128, 133, 135, 142, 146, 165, 169, 193, 206; at airshows, 144; Balkan flight, 136; Communist opposition to, 130, 141; and gliders, 183; as museum piece, 201; Nazi praise of, 175; Polar flight, 148, 166–68; propaganda flights, *175*, *178*, 180–81; reduced activities, 143; Rhineland visits, 138–39, 179, 181, 202; test flight over Germany 1, 130, 163; Vienna visit, 140–41; world flight, 136, 138
Graf Zeppelin II airship (LZ 130), 197, 200–201; construction of, *198*; interior, *200*
Grass, Günther, 209
Great Britain, 72–75, 78, 88, 97, 100. *See also* Zeppelinitis
Gross, Major Hans, 28–29, 31; caricatured, *28*

Hacker, Georg, 185
Hague Convention (1899), 13, 79
Hamburg, 151
Hänlein, Paul, 10, 129
Hapag shipping, 181
Harden, Maximilian, 57
Haussman, Conrad, 101
Hearst newspapers, 121, 163
Hearst, William Randolf, 136
Hedin, Sven, 152–55, 161
Heine, Heinrich, 91
Heinrich, Prince of Prussia, 79, 151, 160
helium, 189, 197–99
Helmholtz, Hermann von, 11
Henson, William Samuel, 12
Hergesell, Hugo, 51, 150–53, 157
Heuss, Theodor, 44, 108, 207
Hindenburg, Paul von, 77, 106, 144, 192, 199
Hindenburg airship (LZ 129), 149, 174; christening issue, 180; construction of, *188*, *190*, 191; destruction of, 172, 182,

195–96; impact of crash, 196; propaganda flight, 180–81
Hindenburg tank regiment, 181
Hitler, Adolf, 144, 174, 199, 206; dislike of airships, 179; at party rallies, 179
Hoernes, Hermann, 52
Hoffmannsthal, Hugo von, 139
hydrogen, 19, 21, 38, 61, 145, 189, 197

ICE train, 208–9
ideology of adventure, 69, 72, 170
I. G. Farben, 165
ILA airshow (Frankfurt, 1909), 63
Inter-Allied Control Commission, 108, 115
International Polar Year, second, 169
Italia airship, 162–63
Italy, 114, 183

Jaray, Paul, 84, 112–13, 118, 189
Jean Paul, 51
Julliot, Henri, 29
Jungdeutschland-Stuttgart, 47
Jünger, Ernst, 203
Junkers *Bremen*, 141
Junkers G-38, 139

Kaiserer, Joseph, 8
Kaiser Wilhelm Society, 160
kitsch. *See* Zeppelin kitsch
Knabbe, Robert, 19
Knorr, 165
Kober, Theodor, 25
Koch, Gustav, 14
Koestler, Arthur, 167–68
Kohl-Larsen, Ludwig, 170
Kölnische Illustrierte Zeitung, 134
Kölnische Zeitung, 136
Kracauer, Siegfried, 145, 193
Krebs, Arthur, 10
Kress, Wilhelm, 52
Krohne (transportation minister), 156, 159, 161
Krupp, Alfred, 127

National Socialism. *See* Nazism
National Zeitung, 44
Naval Airship Division. *See* German navy
Nazism, 146; ambivalence towards airship, 172, 201–2; and education, 183; and modernization, 173; views on technology, 184, 203. *See also* Goebbels, Joseph; Göring, Hermann; Hitler, Adolf
New Realism (*Neue Sachlichkeit*), 111, 132, 144
Nietzsche, Friedrich, 11
Nobel, Alfred, 127
Nobile, Umberto, 161–62
nonrigid airship. *See* dirigible systems; Lebaudy airship; *and* Parseval, August von
Nordstern airship, 114
Northcliff, Lord, 73
Norway, 150
nostalgia, 118–19, 207
Nova Semlya, 167
Nuremberg rally, 178–79

Oppenheim, 37
Ostwald, Wilhelm, 66
Otto, Nikolaus, 17

Paris, 15, 81, 97–98
Parseval, August von, 28, 30, 161; caricatured, 28
passenger amenities, 63, 64, 66, 113, 129, 133, *192*, 193, *200*
Patrie dirigible, 34
Pazaurek, Gustav, 64
Peary, Robert E., 148, 150
Pégoud, Adolphe, 63
Perec, Georges, 209
Pétin, Ernest, 8
Planck, Max, 154
Poe, Edgar Allan, 12
Polar expedition, 125, 147; as patriotic duty, 154, 158, *159*, 164, 166, 168. *See also* Aeroarctic

Popular Front, 183
popular spirit, 40
postcards, as war propaganda, 88, 89, 91, 92, 93, *96*, 100, 105. *See also* Zeppelin kitsch
Poulbot, Francisque, 91
propaganda. *See* Goebbels, Joseph, *Graf Zeppelin* airship; *Hindenburg* airship; Ministry of Propaganda; Nuremberg rally; postcards
propulsion technology, 8–10, 17, 61, 81, 86, 118, 129–30, 184, 207. *See also* Daimler engine; Maybach engine
Proust, Marcel, 99
Pruss, Max, *122*, 195, 202
Prussia: concerns over *Spende*, 125; resentment of, 44, *45*, 47, 56, 118
public: opinion of flying, 4, 11, 24–26, 32, 39, 65; opinion survey, 199; support for Polar airship, 149, 168; support for Zeppelin, 41, 116; understanding of war, 91. *See also* media coverage; youth; *Graf Zeppelin* airships; *Hindenburg* airships

R 34 airship, 116, 122
R 101 airship, 189
Radbruch, Gustav, 69
"rail-Zeppelin," 133
Rathenau, Emil, 44
Rathenau, Walter, 45, 67, 115, 242n. 18
Red Baron, 100
Reichsbank, 136, 196
Reichstag, 32, 40, 101; and air transport, 121; and Zeppelin funding, 34, 78, 80
Renard, Charles, 10, 82;
Revolution of 1919, 112
Robida, Albert, 12
Romain, Jules, 99
Rome, 176
Roosevelt, Franklin D., 199
Rosenberg, Alfred, 184
Rote Fahne, 141

round-the-world flight (1929), 136–38
Royal Aeronautical Society, 182
Royal Swedish Academy, 153

Sachsen airship, 55–56, 64, 140–41
Saloniki, 93
Sammt, Albert, *122*, 181
Samoilowitsch, R. L., 165
Sandt, Emil, 51
Sarrail, General, 93
Saturn V rocket, 208
"Save our Honor" League, 123
Saverne affair, 75
Scandinavia, 150
"Scapa Flow of the Air," 108, 121
Scharnhorst battleship, 180
Scheerbart, Paul, 78
Scherl publishing company, 26
Scherz, Walter, *122*
Schiller, Hans von, *122*
Schlieffen plan, 80
Schulte, Father Paul, 192
Schütte, Johann, 84, 103, 154, 161
Schütte-Lanz, 58, 78, 80, 84, 85, 103,
    154; airship innovations, 82–84;
    patent, 81, 115; use of wood, 82
Schwaben airship, 63
Schwarz airship, *20*
Schwarz, David, 19–20, 28; as hero,
    55, 140
science dividend, of polar expedition,
    167; of space program, 170
science of flight, 7–8
scientific internationalism, 153
semirigid airship. *See* dirigible systems
Shaw, George Bernard, 99
Shenandoah airship, 86
Siegfried, 47, 199
Silberer, Viktor, 52
Simplicissimus, 3, 45–46, 102, 119
social construction of technology, 205
Social Democrats, 41, 52, 61, 69, 119;
    views on Eckener, 145–46
Socialist party. *See* Social Democrats

Society for the Promotion of Aerial Navi-
    gation, 21
Society for the Study of Powered
    Flight, 30
Soden-Fraunhofen, Alfred von, 60,
    62, 81
Soviet Union, 155, 163–64, 167, 183, 201
Spain, 114, 186
Speck, Willy, *122*
speed, of early airships, 10, 17; impact on
    travel, 3; post World War I, 113, 133,
    145; records, 141–42; in World War I,
    85. *See also* airplane, comparison with
    airship; *Hindenburg* airship
Spitzbergen, 151, 153
Stegerwald, Adam, 138
Stephan, Heinrich von, 15, 26, 50, 150
Stockholm, 114, 153, 158
Strasser, Peter, 103, 106, 131
Strategic bombing. *See* airship, bombing
    runs
Stresemann, Gustav, 120–21, 123, 130,
    142; dislike of Hugo Eckener, 160
Sweden, 114, 148, 153, 155
Swissair, 195
Switzerland, 114; reactions to 1908 flight,
    51–52; reactions to ZR 3, 122; tensions
    over *Hindenburg* overflights, 194–96

technological optimism, 33, 69, 111, 120,
    133, 171
technological sublime, 4, 5, 11, 68, 71,
    95, 98, 108, 168, 205, 209. *See also*
    Zeppelin symbolism
technology: displacement of science,
    170–71; distrust of, 24, 39, 67, 111,
    131; Nazi views on, 184, 188, 191,
    202–3
Thompson, Lord, 121
Tirpitz, Alfred von, 106, 184
Tirpitz battleship, 180
Tissandier brothers, 10
Titanic, 208
Tokyo, 136

ZR 3); LZ 127 (see *Graf Zeppelin* airship); LZ 128 project, 189; LZ 129 (see *Hindenburg* airship); LZ 130 (see *Graf Zeppelin II* airship); ZR 3 (LZ 126, *USS Los Angeles*), 112, 128, 132, 146, 152, 156, 160; Communist opposition to, 119; cross-section, *117*; impact on Germany, 116–21, 123–24, 202

Zeppelin Bund (Zeppelin Union), 47, 150

Zeppelin company, 58, 118; averting closure, 114, 124, 152; financial situation, 57, 63, 112, 115, 174, 186; and *Gleichschaltung*, 182; income from charter flights, 144; social unrest, 60–61, 107, 112; subsidies, 155, 174, 186–87. *See also* Colsman, Alfred; DELAG; "Zeppelin-Eckener *Spende*"

"Zeppelin-Eckener *Spende*," 124–26; limited success, 127; use of the polar project, 157–58

Zeppelinism, 41, 136, 171, 197

Zeppelinitis, 72–75

Zeppelin kitsch, 63–64, 94, 178, 206. *See also* postcard; youth

Zeppelin spirit. *See* Zeppelin symbolism

Zeppelin sublime. *See* technological sublime

Zeppelin symbolism, 4, 5, 39, 203; in Imperial era, 41, 45, 48, 67–69; in Nazi era, 172–73, 180, 188, 200–202; in Weimar era, 119, 121, 132–33, 135, 139, 143, 145, 161, 171; in World War I, 80, 99, 206

Zuckmayer, Carl, 37

Zurich, 51